Ruffians & Loose Women

More words derived from textiles

Elinor Kapp

Ruffians & Loose Women
More words derived from textiles

© Elinor Kapp 2016

Published by Elinor Kapp, Cardiff, UK

Distributed by Oxbow Books, 10 Hythe Bridge Street, Oxford, OX1 2EW
and
Casemate Academic, P O Box 511, Oakville, CT 06779, USA

ISBN 978-0-9574759-2-2 paperback
ISBN 978-0-9574759-3-9 epub

The right of Elinor Kapp to be identified as the author of this work has been asserted by her in accordance with the Copyright, Design and Patents Act 1988.

All rights reserved.

A CIP record for this book is available from the British Library

Printed by Lightning Source, UK and Books International, USA

Cover picture designed and embroidered by Louise Gardiner

Illustrations by Jack Coles

Layout and typesetting by Val Lamb of Frabjous Books
www.frabjousbooks.com

... Acknowledgements ...

Many friends have contributed to this tapestry of textile terms.
As with my previous book there are too many to acknowledge them all,
but I particularly want to mention:

My friend David Brown, previously of Oxbow Books, for all his help in editing and publishing this book; also my editor, Jude Irwin, my illustrator, Jack Coles and my indexer, Ingrid Lock.

Dr Robin and Mrs Eleri Gwyndaf. Prof. Charles Muller of Diadem Books.
My nieces Annabel Clemmo and Belinda Kembury and their families.

Amelia Johnstone, Angharad Wynne, Angie Luther, Alison Walker, Anna Whalley, Anne Rahman, Audrey Morgan, Caryl Chambers, Cath Little, Daniel Cohen, Diana Morgan, Dilys Nilsson, Eric Madden, Prof. Ernest Freeman, Frances Hutson in Australia, Gwdihw Curran, Helen Wales, Hiroko Edge, Hugh Lupton, Jacks Lyndon, Janet Cooper, Jan Mullings, Rev. Dr Julie Hopkins, Kaye Edwards, Kay Swancutt, Dr Lata Mathur, Lyn Richards, Margaret and Alan Leisk in Australia, Maris Lyons, Mary and Stephen Ashton, Matthew Harwood, Miriam and Elliot Baron, Peter Williams, Rev. Pauline Warner, Richard Berry, Sally Humble Jackson, Steve Gladwin, Sue Richardson, Steve Killick.

Val Lamb of Frabjous Books for her expert design and typesetting.

... *Dedication* ...

This book is dedicated to my dear family.

To my son Dr Rupert Rawnsley, my daughter in law Erika Rawnsley and the boys, Alexander and Nathaniel, and to my daughter and son in law Amanda and Jon Foster, together with Freddy and Frankie.

Also to the memory of my late husband
Ken Rawnsley CBE

... Introduction ...

Hello my friends, old and new! You have waited a long time for this follow-up to my book **Rigmaroles and Ragamuffins**. Or if you are new to my book, welcome! Wrap yourself in this collection of words and phrases that come originally from CLOTH and have gone on to be used in all sorts of ways in the English language, figuratively and metaphorically.

English is a very rich tongue, with words derived from languages from all over the world and from the beginning of time. Many years ago I was surprised to find how many words came from the many and various types of thread, and my book **Rigmaroles and Ragamuffins** was the result. In it I told of the words and phrases that originated in the raw materials and processes needed to produce THREAD, but had taken wing and become used in common speech as metaphors, similes and even proverbs and folk rhymes.

In this book, **Ruffians and Loose Women**, I continue this theme by looking at the words derived from things made from thread, and of course particularly from CLOTH, not only from cloth for clothing but cloth for sails, domestic furnishings, hats, footwear and many other uses.

Some of these words are still in use in their textile form, but many of them have lost their original connections, like the *nouveau riche* who conceal their humble origin and pretend to be nobly born. So, I am going to 'out' them, but kindly!

Again, I only look at words and phrases that have moved on from purely textile meanings to stand on their own and become metaphors, whether or not they are still used in their original form. It is not a general compendium of textile related words.

Like its predecessor, this book will appeal to all those who love and work with textiles, as well as students of the English language, crossword addicts, translators and linguists. It will be helpful and amusing to foreign language speakers, puzzled and intrigued to find out the origin of some of the quirks they come across in our tongue.

The book is very simply constructed and easy to enjoy. The main entries appear in alphabetical order with a pointing hand indicating related terms that you might like to explore, so that you can browse through subjects that interest you. If you have a particular query, try searching the index! Thanks to its colourful patchwork of snippets of information and short stories, this is the ideal book to dip into between times, when you are following your well-loved textile occupation.

When a word or phrase is being used in its everyday sense, describing textiles, it will be shown in ordinary lower case, like this sentence. CAPITALS highlight words and phrases that are used figuratively or in common sayings, other than when these are part of a quotation. **Bold type** is used to clarify the origin and meaning of words, or to make the tracing of complex derivations easier to follow. Words from other languages are printed in *italics*.

Now read on and prepare to be amused and amazed by the richness that our need for textiles in every part of life has provided. I know you will be pleased to discover why you might put a **Tulip** on your head, and why when you think you're **Vamping** a tune on the piano you may really be mending your socks. Who might be entitled to remove your **Zone**? And what would it mean if you were accused of **Haberdashery** – well you may never have asked these things before, but you will find the answers here!

··· A new tapestry of textile terms ···

ANORAK ☞ JACKET

The word ANORAK was coined in the 1920's and derives from Greenland, where the Inuit people had invented the *anoraq*, a warm hooded jacket made from warm, furry, waterproof sealskin. It became particularly popular for outdoor sports and now can be made in any warm, water-resistant fabric; often a token bit of fake fur lines the hood and frames the face.

In modern times, the word has come to mean **a single-minded bore**, assumed to be **deficient in social skills** and **obsessed with the minutiae** of their pastimes. Calling someone an ANORAK instantly conjures up the image of **a nerdy young man** in the eponymous shabby garment, who is rubbish at chatting up girls. How unfair!

For some reason, train-spotters are often mentioned in this context, maybe because the ANORAK is a cheap and very practical garment, neither chic nor sexy, but good at cutting the wind chill on a breezy railway platform. I couldn't possibly comment, as I can't afford to make enemies of the handsome, strong, intelligent, ingenious, wonderful men we know train-spotters to be.

ANTS IN YOUR PANTS ☞ PANTS

Anyone, particularly a child, who wriggles and squirms a lot might be asked sarcastically, "Do you have ANTS IN YOUR PANTS?" The rhyme makes this phrase more memorable as **a description of restlessness**.

APPLE PIE BED ☞ APPLE PIE ORDER

An APPLE PIE BED, which we children used to construct for fun in the strange old days when humour really was simple, consisted of a bed that looked innocently normal, but had the bottom **sheet turned up short**, so that a sleepy grown-up couldn't get into it. The term may come from *nappe plie,* the French for a **folded sheet**. No doubt the French got fed up with the trick, and that's why they went over to duvets. In America, this trick is known, more prosaically, as a *short-sheeted bed*.

APPLE PIE ORDER

Usually you'll see this in a phrase such as TO PUT THINGS IN APPLE PIE ORDER.

Nappe plie, which as we have seen originally meant a nicely **folded sheet** – probably amid an orderly pile of domestic linen – may have given rise to this term. People who couldn't understand French, or perhaps disliked the language as a 'foreign' imposition, garbled the words. Alternatively, the words could be a corruption of the French *cap à pied* suggesting something well ordered **from top to toe**.

APRON ☞ PINAFORE, BIB AND TUCKER, and BRAT

Another word for an apron is a pinafore, which tells you exactly what it is: something that you **pinned afore** or **in front** of you. Apron was originally a *napron*, a **tablecloth**, so it has lost its **n** and remained as a textile term basically for some form of **protective covering for the body**, especially **in front**. However it also gave us a metaphoric use as the APRON of a theatrical stage or in an aircraft hangar. We still have vestiges of the old tablecloth connection in the word NAPERY which denotes **household linens** such as table coverings and napkins.

The phrase TIED TO A WOMAN'S APRON STRINGS is used slightly scornfully to suggest **a man who is overly dependent on a woman**, as in the bluesy song verse, "St Louis woman, with all her diamond rings, hauls her man around BY HER APRON STRINGS".

ARMOUR ☞ COAT and SUIT

The word ARMOUR as a description of the big clanky **metal protection** worn by knights and other warriors, has become a constant sight in art, book illustrations and on our TV screens. We may romanticise "a knight in shining armour", but it's almost impossible to imagine the reality of riding and fighting in chain mail and ponderous, jointed steel panels, though there are plenty of re-enactment societies to try to show us. The word pretty obviously originates in *arm* for the upper limb, and hence **weapons and coverings of these appendages**, and has a Latin root.

In the Middle Ages, there were also COATS OF MAIL; made of **linked loops of steel** to protect against dagger thrusts and spear wounds.

I'm not really sure if ARMOUR or A SUIT OF ARMS quite qualifies for inclusion in a book about textiles; most of us would prefer soft silks and satins to rusty tin and harsh iron for clothing! However, we speak of being ARMOURED AGAINST a number of things metaphorically, meaning that **we try to protect ourselves** against TROUBLE, BAD NEWS, EMBARRASSMENT or really any unpleasantness, using the metaphor of wearing ARMOUR as **protection in battle**.

A COAT OF ARMS is more clearly textile related, as it began as **a silk tunic embroidered with the heraldic insignia of a feudal family or institution**. Symbols such as noble birds of prey, stags, plumed helmets, chevrons and shields sent a sort of shorthand message about where your loyalties lay that could be 'read' even by illiterate common soldiers. Being able to identify where your troops should rally and which lord was in charge was rather important in more belligerent days, and very handy amid the furore of a battlefield.

Today, such insignia of nobility or connections with important forebears are mostly ceremonial and are strictly regulated by the College of Heralds – rather like registering a modern commercial brand name. So it's no good you concocting a wild and wonderful coat of arms for yourself and expecting posh restaurants to give you their best tables in deference to your important ancestors. The Heralds have a whole traditional vocabulary for every colour and device, and rules that make fascinating reading for those who love arcane information. In medieval times, when a marriage took place, the intricate COAT OF ARMS of each noble party would be cleverly combined. Just imagine how busy Henry VIIIth must have kept those Heralds!

ASBESTOS

If you can drink your coffee boiling, or eat a Clarksie straight out of the oven you might be told admiringly that you have **an ASBESTOS throat or ASBESTOS hands**. (Please be advised that where I come from in Wales, UK, the men are Real Men and a Clarksie is a meat pie).

ASBESTOS has been mined for over 4000 years, and many cultures, including the ancient Romans, were aware of the surprising ability of this strange, fibrous mineral to resist heat and revered it for its inflammable qualities. Its name comes from a Greek word meaning *inextinguishable*. Although ASBESTOS comes out of a lump that looks like a rock, its tough, matted crystals were recognised and used as a textile from earliest times.

When King Charlemagne held feasts, he spread a tablecloth for his guests which was made of ASBESTOS. At the end of the meal, the servants would dramatically cast the whole cloth, food mess and all, into the fire, from which it would emerge clean and shining. Hooray – no more washing!

It is also said that in ancient China the nobility liked to wear long, draping sleeves to their gowns, but these dragged over the food and got very messy. Solution? They had the weavers make detachable sleeves of ASBESTOS which could be burned clean after dinner. I have been unable to verify this story and doubt is cast on it by archaeological analysis of linen costumes. It has all the hallmarks of what we now call 'urban legends', but I enjoy preserving these tidbits of folklore too, so long as we don't take them too seriously, or fight over them.

The disadvantage, of course, is that ASBESTOS can be highly dangerous to health if the fibres are inhaled. Even the Romans knew it had toxic effects on the lungs, and Pliny the Elder noted that those who mined ASBESTOS or wove it into fabric died early. It is shocking that it was not until well into the 20th century that workers exposed to it were heavily protected, and domestic asbestos was progressively removed wherever it was found. I met an elderly lady whose mother and great aunts had been asbestos spinners in the north of England, and all had sad tales of the lung conditions and early deaths their families had suffered.

BAG ☞ BAG LADY, PURSE, SACK

Purses and bags are an essential for carrying things, and have been made out of every possible material, from a whole small animal's hide, to workers' sacks made of coarse hemp and canvas, to the most delicate purse that a lady could net, sew, knit or crochet in her own drawing room. It was customary to make such utilitarian items for men, too, since they had just as much need to carry money and other small items, long before the days of smart leather wallets.

LETTING THE CAT OUT OF THE BAG tends to be used now in the sense of **telling something that should have been kept secret** – or, at least, revealing it too soon.

BAGS I! is schoolboy slang **to claim** something, perhaps from the idea of capturing something in a bag, though the origin seems to be unknown and it is rather outdated slang. The term DIBS, or I DIBS IT! can also be used for this.

IT'S IN THE BAG probably means a hunter's bag to carry game, and figuratively you say it when **you are absolutely confident that your plans will turn out right**. An alternative explanation is that it relates to petitions brought before the English Parliament and put into a velvet bag as a batch to be considered. A bag still hangs on the Speaker's chair, but it is now purely symbolic.

To be LEFT HOLDING THE BAG is a way of saying **you have been made to look foolish or lumbered with unwanted responsibilities**. Various reasons have been given for the term. One of the more colourful came from rural America, where a gullible youngster would be persuaded to hold a large bag open by bushes to catch non-existent birds, while the prankster ambled off on the excuse of flushing out the game. It might take quite a while before it dawned on the victim that he'd been had. There is a long tradition of playing tricks on new apprentices, such as sending them off on plausible but absurd errands like this.

A Celtic story uses exaggeration to good effect in the tale of Wee Jack who wins a lying contest and becomes King. His lies become more and more absurd – especially his claim to be carrying 500,000 tons of grain in a BAG made from the skin of a flea. Nowadays, it is mostly politicians who solicit our votes and get into power by telling more and more barefaced and outrageous lies.

We surely must not ignore the MANBAG, although this word has really only caught on in the last few years. These words, often coined by journalists, try to avoid sexism by adding the word Man before them – the MANKINI for **a male version of the bikini** [qv] is probably the most horrible example, but the derogatory 'Man flu' runs it close. A MAN BAG is, just as it sounds, a **male version of the handbag** many Western women carry around to hold useful personal items. Such words can be described as highly egalitarian – they annoy both men and women equally!

BAG LADY

A BAG is also **an uncomplimentary name for an old woman**, as is A BAG LADY, a **female tramp** carrying all her possessions in carrier bags. A young and attractive woman might be called a BAGGAGE affectionately nowadays, suggesting that she's **a saucy girl**, or **a bit of a handful**, rather than with the earlier, derogatory meaning of **a loose woman**.

BAG AND BAGGAGE, however, has nothing to do with women, saucy or otherwise. It started life as military jargon for all a soldier's gear, and TO GET OUT, BAG AND BAGGAGE meant **the army was retreating with honour** without having to surrender their possessions. However, in civilian life, this phrase has come to mean **slung out on your ear with everything you own**. So just make sure you pay your rent on time!

BAMBOOZLE ☞ COLOURS

Imagine sailing on the ocean near another ship that's flying the flag of your country. Suddenly, it starts firing its guns to sink you, because it's really part of an enemy fleet, and is only using its friendly flag as a disguise. You've been BAMBOOZLED.

The word BAMBOOZLE now means simply **to hoax** or **confuse someone**, but it may have arisen in the British Navy between the late 18th and mid 19th centuries, with the specific meaning of **to deceive the enemy by sailing under a false flag**.

This is such a nice story I much regret that honesty compels me to say it is probably an invention, and that we have no evidence of origins earlier than the 20th century. Proper Lexicographers can argue about such things for hours, which is probably their idea of a fun day out. One source did suggest a link with bombast, which comes from the Greek word for silk. Unless you accept that equally dubious idea, I really shouldn't have included the word in this textile-related book. Thus I have, in fact, BAMBOOZLED you. So there!

BANDANNA

A BANDANNA is a **large colourful handkerchief**, often used as a neck scarf in casual male clothing, perhaps in imitation of gypsies. It sounds as if it ought to come from band or bandage but in fact was a Hindi word for a type of tie-dying, *bandana*, which gave a pattern of white or yellow spots on a coloured silk ground.

There was even a 19th century slang word for a thief who specialised in stealing BANDANNA handkerchiefs, an ABANDANNAD, perhaps a combination of abandoned with bandanna.

BANDBOX ☞ HATS, RUFFIAN

To comment that someone looks as though he had just STEPPED OUT OF A BAND BOX is to say that he is **most fashionably dressed** and **meticulously neat**, and, by extension, that he is quite a dapper fellow. Indeed, it can sometimes imply that he is positively dandified, and therefore a bit in love with himself.

It would be quite hard to step out of a BANDBOX in reality, as it is a small container made out of light cardboard. In other words, a HAT BOX. Nowadays, some of us would store delicate HATS or Fascinators in such a container. Fascinators of course, for those not in the know, are those tiny frivolous head decorations made of lace, feathers and sparkly things, with the intention of fascinating men. They may indeed do this, but only until the said male notices the price tag!

In the early 17th century, the term BAND BOX was coined for a box in which the BANDS or RUFFS of both men and women were stored or carried about. Elaborately-ruffled collars needed protection in a container, as they were often made of lace, stiffly starched and a foot or more in diameter. Later, as fashion changed, similar boxes were used to protect equally delicate hats, celluloid collars, and the like.

··· *Bandbox* ···

So, I'm sorry to disappoint you, but the BANDBOX in the saying is probably nothing at all to do with a small gazebo in the park where a musical BAND are playing – even if the musicians are all very natty chaps.

BANNER

A banner is a long narrow flag, a smaller version of the flag called a 'Standard', much used in Heraldry, and its name comes from an old Norse word *benda*, to **give a sign**, which lost its **d** and became the Anglo-Norman *baner*.

A BANNER is therefore a strip of cloth held high, to make **a statement about your allegiance**. Even more than other flags, it displays information about your **clan, family** or **affiliation**, such as the Star-Spangled Banner or American flag, and the Union Jack of Britain.

We might be figuratively UNDER A BANNER to suggest **a cause to which we are devoted**. As for me, I'M CARRYING A BANNER FOR THE PLACE OF TEXTILES IN OUR CULTURE – especially their place in our language.

BASTARD

Bast is a tough fibrous material obtained from the inner bark of plants, such as jute, hemp, flax and the Linden tree. One of its uses was as stuffing for hardwearing articles such as mattresses and pack saddles. Indeed, the Latin name for a pack saddle was *bastum*. So how did the word BASTARD for a **child born out of wedlock** come about? Such children in France were once known as *fils de bast*, literally *son of the pack saddle*, the idea being that it was often mule drivers, soldiers and other fly-by night itinerants who would sleep one night in a village using the packsaddle as a pillow, and they would be blamed (rightly or wrongly) for the babies born in due time.

When we consider that the fibrous cellulose material is derived from wood, it is not surprising that the word *bat* has been used historically both for various *sticks* and for *felted, tough padding* – although the latter is often spelt *batt* or *batting*. The word BASTE is of similar origin, meaning **to mend** or **patch**, as well as **to sew together temporarily with long stitches**. It probably has links to a lost Germanic word for binding, as in a *fillet for binding the hair*, and even to an old Irish word, *basc*, meaning **necklace**.

Given that this type of padding is tough and hard it is perhaps not surprising that the word has become linked to other irregular items. The term BASTARD SWORD could refer to a particular type of **long or heavy sword**, or to describe a sword that is **in between two other known types in weight**.

The term has also been used as an 'intensifier', as in the case of other vulgar expletives, to strengthen the sense of masculine power.

Other tools are described in the same way. From the start of the 20th century, the term has been used for BASTARD FILES, which are **files of a specific degree of coarseness.** This is no doubt something about which file devotees can happily argue for hours. I doubt if I could add anything to which they would listen, except to point out that FILE also owes its origin to a textile-based practice – that of stringing parchment documents across strong threads, or *fils* by medieval Monks to let them dry without smudging.

{*For more details about the word FILE and also BASTE, BAND and BIND, see my first book,* **Rigmaroles and Ragamuffins**}

In the past, calling someone a BASTARD made it clear that he or she was not a legitimate heir to a family's assets; Kings and other nobles often sired such children 'on the wrong side of the blanket' (see BLANKET) and in the English name Fitzroy the *fitz* denoted illegitimate child, while the *roy* was the king. BASTARD then was an insult about someone's parentage, but today, may simply denote a person who makes your life difficult, or be used jocularly or in a matey way with a friend you know well.

The word BASTARD with a change of initial letter to an 'f' is probably also linked to *fasces,* derived from the Etruscan *fasces* – **a bundle of sticks**, sometimes with an axe protruding in the middle, which the ancient Romans – and much later, the 20th century Italian **Fascist Party**, adopted as a symbol of authority. The underlying idea was that you could break one rod, but bound together, they were too strong to break.

FASCIA also describes **sheets of strong, fibrous tissue** in the muscles or beneath the skin. This word pops up again as an architectural term for **a board that covers the ends of joists or finishes off edges**. A FASCIA BOARD **is also the backing upon which business signs are attached** above their shop's street frontage.

BED ☞ BLANKET, CURTAINS, DUVET, SACK and SHEET

Tracing the origins and relationships of words can be quite tiring, and already we could do with a little rest along our way. Let's take it easy as we look at some of the metaphors and figures of speech surrounding that blissful – for many reasons – place we call BED and the textiles associated with it.

The word probably sprang from an ancient Indo-European source, now lost, which combined the idea of **somewhere to sleep** with **somewhere plants grow**. Mankind's original sleeping places were perhaps no more than shallow holes scraped in the ground, like an animal's lair, as suggested by the Roman word *fodere* to *dig*, which (with the **f** transposed to a **b**) eventually gave us our BED.

We still use the same word for the place where both people and plants can stretch out their toes. Occasionally, both sorts of living things may sleep in the same garden bed, if they are the sort who habitually come home drunk and fall over into the herbaceous border. I mean people, of course; plants are much better behaved.

There are quite a few phrases which deal with going to bed, and – not surprisingly – most of them can be used both in a straightforward way, or act as euphemisms for having sex. If you are not fluent in the English language, take care which phrase you choose! To say "ARE YOU GOING TO HIT THE SACK?" or "I'M GOING TO SACK OUT NOW" would be fine to tell someone that you are just **going off to bed to sleep**, but talking about BEDDING SOMEONE or GETTING THEM INTO THE SACK suggests **sexual seduction**.

TO FEATHERBED SOMEONE is **to give them a nice cushy life**, perhaps excessively so. It often carries a note of disapproval, as in, "I'M NOT GOING TO FEATHERBED HER ANY LONGER".

However, someone might still shake their head and warn that **you will have to live with the consequences of your mistakes**, in the proverb, AS YOU HAVE MADE YOUR BED, SO YOU MUST LIE ON IT. Indeed, your **bedfellows** may not be literally in the same place as you, but metaphorically applies to your cronies or partners in any enterprise.

A children's sleepy-time rhyme from Bedfordshire, a county in the UK, shows, with creepy-crawly finger movements, how you must go:

> "Up the Wooden hill to Bedfordshire
> Down Sheet Lane to Blanket fair."

BELT ☞ ZONE

Tailors are not generally known as fearless fighters, and are associated with making clothes rather than embroidering. However, in one Grimm's story lampooning the absurdity of the traditional heroic stereotype, a tailor embroiders the words "Seven at one Blow" onto his belt. This refers to his having killed seven flies with one thwack of a rag, but everyone around him believes it means men. As a result of this misconception and his quick wits, he earns popular respect, makes good in life, eventually kills a giant and wins the hand of a princess in marriage. He certainly didn't suffer for EMBROIDERING HIS STORY.

BELT has a surprisingly uncertain origin for such a basic piece of clothing. The Romans used *balteus* that could have been borrowed from an Etruscan word, now lost.

To say to someone, BELT UP! meaning **shut up** seems to have originated just before the Second World War, and was mildly rude when I was young; we would certainly be ticked off for using it. Presumably to BELT someone originally meant **to beat them with a belt** and was also slightly vulgar, at least for girls, mainly because you were not supposed to do anything so vigorous as hit someone. Sly use of social ostracism, lying, gossip and poisoned compliments were once the weapons of choice among the more ruthless sort of schoolgirl. Are they still? Hmmm. Let us pass on quickly to another topic.

You can also BELT OUT A TUNE, by **singing forcefully** and **loudly**. From that came a more modern usage as A BELTER, meaning **something exceptionally good** of its class. THAT WAS A REAL BELTER OF A TUNE, for a performance, or even OF A GOAL if it has been **kicked in very hard**. It can be more or less anything that was **excellent and praiseworthy**, particularly where effort has been used.

A BLACK BELT is a **top judo expert**, the different levels of expertise in this martial art are shown by the different colours of fabric belts worn by contestants. HITTING BELOW THE BELT is a boxing term. The Marquess of Queensbury rules, laid down in 1865, prohibit hitting the solar plexus or the groin – anywhere, in fact, below the belt, as being both dangerous and unsporting. Today, the term means **to act unfairly in any way**.

We use the term BELT AND BRACES to mean **taking every possible care** to prevent disaster. TO TIGHTEN YOUR BELT means **to spend less, to economise**, presumably because you get thinner, and GETTING IT UNDER YOUR BELT means **you have got something or mastered a task**.

BIB AND TUCKER ☞ APRON, BRAT and PINAFORE

"PUT ON YOUR BEST BIB AND TUCKER and I'll take you to the Ritz!" someone might say to me. Well, a girl can always dream.

The phrase dates from the 17th century, when men and women alike would try and protect their richly-brocaded clothing from their messy eating habits. Only linens and cottons had any chance of being washed, and that not too often.

A woman wore a piece of lace or linen known as a tucker round her neck, which also filled in her low neckline for modesty. A man sported a bib, a shaped cloth tied around his neck. The word is likely to have come from the Latin *bibere,* meaning to drink, and is still found in words like imbibe, bibulous and bibacious. The word BIB is also used for **the top square of an apron**, worn by either sex. Together a couple could be going out to feast on a holiday dressed in their BEST BIB AND TUCKER.

The term may have fallen out of use in the UK but from 18th century Lancashire, where it was used for women's wear, but rarely that of men or children, it leapt the Atlantic to the USA and then was re-imported to Britain around a hundred years later. The term TO NAP YOUR BIB, current in the 18th century but now obsolete, meant **to cry** – or more literally, **to use one's bib to wipe away tears**.

In the US the term BIBS arose as slang for overalls, also called 'dungarees'. Fascinatingly, the latter name came from a Hindi word for a sort of coarse calico made near Mumbai that became known to Westerners around 1610. Heaven knows how it survived and transmogrified, but that's a different story. For those not familiar with BIBS, they are trousers and top in one, with a squarish bib at the front and broad shoulder straps that cross at the back, made of a strong serviceable material such as denim and worn with a shirt beneath. In the deep South or "good ol' boy" territory, you may glimpse portly gents wearing bibs shirtless, with ample flesh spilling from the edges. Not a pretty sight, really. The equivalent garment for women is the PINAFORE DRESS which, I may add, is always worn with a blouse or jumper underneath.

A child's BIB is a cloth tucked or tied round the infant's neck to catch the products of dribbling, bubble-blowing, spitting, smearing, self-inundation or possetting – baby habits you will find endearing or disgusting, depending where along the spectrum of mother-love you are at that particular moment.

The earliest use of the word BIB was in a dictionary of 1580, as meaning 'for a child's breast.'

In our family, at least, the word TUCKER was also used for a child's bib. I haven't traced this in the literature; it could have been dialect, or simply an idiosyncratic use. To this day, I have a beautifully embroidered double bib, worn by my father as a baby in the late 19th century, and always described by him as a 'tucker'. One side was semicircular and used when eating, and my father told me that after the meal his mother would turn it round so that the square embroidered tucker was on show, unsoiled, at the front. So, to me as a child, WEARING YOUR BEST BIB AND TUCKER was an ironic phrase for a baby's garment.

TUCK is an interesting word. Its source is an Indo-European root word conjectured as *deuk,* meaning to pull, from which are derived conduct, duke, educate, reduce, tie and tow. The Germanic base *teuk* became English TUCK to denote **pulling up** or **gathering**.

Since **tucking** is a process of gathering, folding or pleating cloth that makes it smaller, it probably gave rise to the idea of TUCKING something into a small space or crack **to conceal it**, just as TUCKING the edge of a sheet or other drapery beneath something holds it in place.

By the late 18th century, the word was used as slang for **devouring food or drink, especially in a greedy or hearty manner**. This may derive either from TUCKING the food into the crack in your face, or from TUCKING a napkin into your collar or dress before eating. Master etymologist Partridge has yet another suggestion. An early expression for **a large meal** – similar to our modern slang "a blow-out dinner" – was TUCK OUT, when one obviously removed the tuck or crease from one's waistband. We still TUCK IN TO A GOOD MEAL, and the Australians retained the 19th century word TUCKER, originally the **daily rations** of a gold prospector or farm worker, and now use it to mean **any food**.

From the mid-19th century, British public school boys (which, confusingly, actually means private school boys) usually had **extra food or special goodies** – their TUCK – sent from home or bought it in the TUCK SHOP attached to the schools.

TUCKERED OUT, often uttered in corny American Westerns as I'M JUST PLUMB TUCKERED OUT, is actually still used to mean **exhausted** in the USA. It was probably derived from the same source as the now obsolete TUCKED OUT in the UK, and probably referred to the **appearance of an overdriven horse**, with its ribs showing like tucks in the skin as it heaved for breath. Veterinarians sometimes speak of animals that are TUCKED UP, meaning they are **holding their abdomens in an unnaturally tight position**, which suggests they are **in pain**.

It's much nicer to be TUCKED UP in bed by nine o'clock with a hot cup of cocoa, which simply means you are warm and cosy under the duvet or blankets. However, if you have been **cheated** or **deceived** by someone in the UK, you might say, "I'VE BEEN TUCKED UP" (or more colourfully, "I've been done up like a kipper.") Getting messed about is bad enough, but you certainly don't want to be TUCKED UP in the 19th century sense, because dear Mr Partridge says it meant **to be hanged!**

BIKINI ☞ BAG

This garment is occasionally classified as underwear, but was designed to be worn on its own as feminine beachwear. The BIKINI was named after an atoll in the Western Pacific Ocean, the word BIKINI – originally *pikini* – meant, in the Marshall Islands dialect, *coconut place,* but these 23 dots of land had the sad distinction of being the site for the first post-war nuclear bomb tests in 1946. Just like the eponymous tiny two-piece bathing costume, the atoll could hardly be seen, but the radioactive pollution proved to be a very big problem. To this day, none of the evacuated families has ever returned to their homeland. The fashionable bathing suit's name – wouldn't you know it? – was coined by the French, appearing first in their news magazine, *Le Monde Illustré* in August 1947.

Some of us still remember (though probably wish we could forget) the irritating 1960 pop song about AN ITSY BITSY TEENY WEENY YELLOW POLKA DOT BIKINI, and could even hum along to Bryan Hyland's hit song.

For a peep at the origin of the horrible MANKINI word, see BAG.

BIRTHDAY SUIT ☞ EMPEROR'S NEW CLOTHES and NAKEDNESS

In the Bible, we read the story of Adam, Eve, and the apple. We are told that, when the first two people realised they were naked – after eating the forbidden fruit of the tree of knowledge – they sewed fig leaves together and made themselves the first ever clothes. But wearing nothing at all is how we are born, and **nakedness** is described metaphorically as BEING IN YOUR BIRTHDAY SUIT, a phrase first noted in the 18th century.

BLANKET

As with the word blank, the word BLANKET meant originally a piece of **white cloth**, as it came from the Roman *blancus* for white, and by way of Old French *blancquet*. BLANKET denoting a **woollen covering** has been in use since at least the 15th century.

TO LIE IN THE WOOLLEN was **to have a blanket next to your skin**. Prickly. "I could not endure a husband with a beard on his face, I had rather lie in the woollen," says Beatrice, the feisty heroine of William Shakespeare's 'Much Ado About Nothing'.

To be buried wrapped in a woollen shroud was at one time required by a law enacted in 1668 under Charles II, King of England. It was intended to protect the domestic woollen industry, which had been in decline. Only extreme poverty and plague were excuses. If you used a fabric made in foreign lands for a shroud, the fine was £5, a huge sum for ordinary folk then. The law was repealed about 11 years later.

There is a touching story about a young American Indian girl who had a rainbow cloak which was so beautiful it was stolen by the sun, who threw it around his shoulders in the morning and evening, producing the wonderful colours of sunrise and sunset.

A similar story was told about a mischievous boy from the indigenous Thompson River people of British Columbia. The lad wandered away from home, and on returning, found that his parents had moved away. His old grandmother had been left behind, and she taught him how to make a bow and arrows to supply them with food. She made blankets for him out of the skins of many-coloured birds. The feathers were especially fine from the blue jay, as were the reds, yellows and oranges of the other birds. The Sun saw the blankets, and found them so beautiful that he bought them to wear, having previously only covered himself up at night and gone naked in the day time. He wrapped himself in the garments and disappeared into the sky. Ever since we glimpse his wonderful robes in the evening after he moves beyond the horizon.

Maybe the first blanket was devised in prehistoric times by a cavewoman who got fed up with her mate hogging the mammoth hide at night. Perhaps she idly began to pull out bits of woolly fur and twisted them together. She had invented yarn, and during cold, wakeful nights she began weaving the fibres in and out of each other on a simple framework of sticks and sinews. Eventually, she finished her task, wrapped herself up snug and warm, in the world's first woolly blanket, and went to sleep on the other side of the cave (divorce not having been invented), leaving the old boy snoring under a hard, bare, smelly piece of leather.

BLANKET DRILL was slang from the British Army in India for **a siesta**. We can BLANKET THE EARTH WITH SNOW or even with SILENCE, and TO BLANKET A SOUND means **to muffle it**, as if with a blanket. A BLANKET PROHIBITION **completely forbids something**. A WET BLANKET is **a spoilsport**, and WHY BURN THE BLANKET TO GET RID OF THE FLEAS? is a readily understandable proverb from Turkey.

BORN ON THE WRONG SIDE OF THE BLANKET was quite a picturesque way of describing **someone who was born out of wedlock**; that is, to an unmarried woman. The term was first noted in the latter half of the 18th century, but did not come into general colloquial use until the middle of the 19th century. The phrase is not much heard today, because the stigma of illegitimacy is no longer attached to the innocent child.

One idea that is almost certainly as old as textiles and human families, is that a soft bit of old cloth can be given to a baby, to console it, and perhaps carried and cherished by the child for years after. **An item that eases us from one age into another** in psychoanalytic jargon is a 'transitional object'; but how much nicer to call that dear old rag a SECURITY BLANKET, a term apparently coined by Charles Schulz, creator of the cartoon strip, 'Peanuts'. It was the BLANKET that Charlie Brown's thumb-sucking friend, Linus, trailed around all the time in his wake. The phrase caught on and is now used to mean anything that serves to prop up our insecure little egos in fraught and anxious times. In passing, please note the link of LINUS the name, to the Latin, *linum* for a thread!

BLOOMERS

The word BLOOMERS comes from the great dress reforms of the mid-19th century. Mrs Amelia Jenks Bloomer, a devout Quaker and Temperance advocate, recognised that women's tightly-laced corsets posed a serious health threat. She set about popularising a more sensible costume, designed by a Mrs Elizabeth Smith Miller of New York; thereafter, BLOOMERISM came to stand for all **rational dress reform**. In form, BLOOMERS derived from baggy Turkish trousers, held in at the ankles.

They were worn under a full skirt – often short enough to reveal their ornate laciness. Unfortunately, they looked a little too much like trousers, and some of the men – and women who liked to be under their thumbs – felt threatened by that resemblance. BLOOMERS deserved a lot better press than the ridicule they attracted, and were a practical and decent variation on women's dress.

The term TO MAKE A BLOOMER meaning **a mistake** is of uncertain origin, but dates to the late 19th century and was probably Australian.

BLOUSE

Women today wear white or coloured BLOUSES with skirts almost as a working uniform. It is interesting to note just how far back this garb goes. The Greek goddess Hera, wife of Zeus, is described as putting on a *heanon* when she visited her husband. Appropriately, this forerunner of the vest as warm underwear was given her by Hestia, goddess of the hearth and home, who keeps you warm. But the invention of a BLOUSE may have been even earlier, as it appears that Bronze Age women wore only three garments: a white tunic, a belt, and an oblong or tubular overwrap. That's really not so different from today's white BLOUSE, skirt and belt.

In the northern counties of Britain, where men pride themselves on their masculinity, **any man who doesn't match up to the prevailing macho level** may be told he is A BIG GIRL'S BLOUSE. Although it is meant contemptuously, it is usually said in quite a good-humoured way. This may be because it was introduced into common usage in the 1960's, an era of popular sitcoms from the North, and it is also sometimes rendered as A BIG GIRL'S SHIRT, a variation from America, where the original was not well known.

Incidentally, BLOWZY sounds as if it should be spelt BLOUSY, resembling a textile word, since it certainly conjures up a buxom wench with her bosom spilling out of her blouse. However, the word is of unknown origin and more likely linked with 'blow', in the sense of the blossoming and blooming of flowers. This full or even over-blown sense fittingly described a disreputable, **low-class woman**, often of a ruddy or dissipated complexion.

BLUE RIBBON

By contrast with the previous specimen, to belong to the BLUE RIBBON ARMY used to mean **someone who never drank alcohol**, though it is rarely used now except as a jokey expression. In the 19th and early 20th centuries, drunkenness was a huge private and public problem. The Blue Ribbon Army started in the USA and extended to the UK by 1878. The members wore a small blue ribbon as a badge and signed a pledge never to drink alcohol.

However, describing someone as A BLUE RIBBON means they have been **invested with The Order of the Garter**, an honour from the Queen of England. Because it is the highest honour she can bestow, it is used metaphorically of **the 'top' job or honour** in other fields. Horses may earn a BLUE RIBAND by coming first in certain races, and the term also means an unofficial accolade for the fastest ocean liners crossing the Atlantic.

BLUE SKY

Is there ENOUGH BLUE SKY TO MAKE A DUTCHMAN A PAIR OF TROUSERS? we used to say, looking hopefully up on a wet day when the rain clouds were beginning to blow away. Some people might say it as, ENOUGH BLUE SKY TO MAKE A SAILOR A PAIR OF TROUSERS. Either way, these trousers are supposedly rather large, and usually blue in colour.

One wonders what the Dutch themselves might say, and apparently there are different versions in the Netherlands, too. ENOUGH BLUE SKY TO MAKE A PEASANT'S SMOCK, or TROUSERS FOR A GENDARME or A CUSTOM'S OFFICER.

The most charming saying comes from the Walloon district of Belgium, where you look skyward to see if there is ENOUGH BLUE TO MAKE A CLOAK FOR THE VIRGIN MARY AND A PAIR OF STOCKINGS FOR BABY JESUS.

BLUE SKY THINKING is one of those annoyingly meaningless pieces of management-speak, used to suggest **clear and excitingly large possibilities**. Oddly enough, it apparently sprang up in the early 20th century to suggest **fraudulent and empty speculation**. Perhaps it should have stayed there, since there is still so much of that chicanery about.

BODICE-RIPPER ☞ LOOSE WOMEN

The original form of the corset was a pair of 'bodys'; two pieces, front and back, made of shaped and stiffened linen, laced together at the sides. This later became known as a BODICE.

A BODICE-RIPPER is the name in the book trade for **a steamy novel set in historical times**, since at some point the hero was expected to tear off the heroine's clothes in a grand seduction scene. It is comparatively modern, being recorded first in the 1970's.

BONNET

The word BONNET probably came from an obscure Latin word *bonetus*, which became *bonet* in Old French, meaning the **fabric used to make any type of soft headwear**. Not only women wore bonnets; it is still the term for **a Highland Scot's distinctive cap**. Various **protective coverings** for machinery can be known as BONNETS for obvious reasons. In a car, **the rounded bit at the front** where [mostly] the engine is housed is what Americans call the 'hood'; it is known as the 'bonnet' in the UK.

A BONNET is also a nautical term for a sort of 'add-on' **sail that is laced to the bottom of a larger sail to catch more wind.**

If you THROW YOUR BONNET OVER THE WINDMILL, you are figuratively **doing something quite reckless**. However, if you plan to WEAR YOUR EASTER BONNET you are generally in a springtime mood, as if **dressing in your best clothes and a new hat** for church on Easter Sunday.

To say YOU HAVE A BEE IN YOUR BONNET means you are a bit **crazy about a certain subject** or **you can't stop talking about it**. You might well be preoccupied in those circumstances, as a **bonnet** is also the name of the protective headgear worn by beekeepers, which is intended to keep the bees out.

There is an English folktale from Flamborough, Yorkshire about a water spirit who lived in a deep pool. The spirit that haunts it once belonged to a young woman called Jenny, thought to have drowned herself there. If you run round the hole nine times, widdershins [against the sun] her spirit will rise up and drag you in, saying,

> "Ah'll tee on my bonnet,
> An' put on my shoe,
> An if thee's nut off,
> Ah'll soon catch thoo."

A farmer who got a bit tight at the fair, boasted he would do this, rode his horse round the mere nine times, and only just got away when Jenny rose out of the spring. She failed to catch him, but bit a piece out of his horse's flank, which showed a white patch ever after.

BOOT ☞ ROBE and SHOE

The word for a Boot, usually meaning a solid shoe with an ankle support or occasionally, a sheath for the leg, probably came from the Old French *bote,* but its earlier forms are unknown. Coverings for the foot and lower leg made of leather or other fabric may have been named for BOOT meaning compensation goods. Alternatively, it can relate to mending or making something right.

The word was also linked to BOOTY denoting spoils of war. I somehow feel that even my best tall, shiny boots, much as I love them, would not quite equate to either of those definitions, though I do like the idea of a ravening horde who ran away with the boots but left everything else in my house – including me – untouched.

The word BOOTY also has a link to FREEBOOTER, for a pirate or other robber who plunders ships. It comes from the Dutch *vriebuter*, so the original link is with ROBE for a dress [qv] via 'robbed'.

This early use of putting things right is still found when we speak of REBOOTING a computer, which [I am told] is not really the swift kick I frequently aim at mine in exasperation before falling over and bursting into frustrated tears, but simply means **switching it off and on again** to restore it.

We still speak of BOOTING someone out, or PUTTING THE BOOT IN, which can be used literally as stamping on or kicking someone [please don't!] but more commonly suggests giving someone **a good telling off** or other punishment. Similarly, we use it metaphorically when we say, you might BOOT ME OUT, or

GIVE ME THE ORDER OF THE BOOT, clearly suggesting you want to **get rid** of me.

The phrase TO BOOT can also mean **in addition to**. For example, "that car driver gave me a rude sign as he cut in, and he was driving much too fast TO BOOT". However, it actually comes from a different and non textile source, the proto-Germanic *boto* and the Old English to *bote*, from which we also get 'better.'

BOOTLESS is also an archaic term for **useless** or **fruitless**. One of the most memorable and poetic employments of this last sense can be found in a Shakespearean sonnet.

> "When in disgrace with fortune and men's eyes
> I all alone beweep my outcast state,
> And trouble deaf heaven with my bootless cries,
> And look upon myself, and curse my fate....."

Poor old William was describing a situation when HIS HEART WAS IN HIS BOOTS. This did not mean that he had somehow suffered cardiac slippage and would have to shake his heart out of his footwear like a pebble and pop it back in its proper place – he was simply feeling **lost in utmost sadness and desolation**.

If I'd become deeply ashamed of something I'd done, or were an extremely subservient sort of person (which I am not!) I might offer TO LICK YOUR BOOTS in a cringe-making act of **self-abasement**. If you allowed or encouraged me to do such a thing, someone would undoubtedly accuse you of BEING TOO BIG FOR YOUR BOOTS, meaning you were **puffed up**, **proud** and **considered yourself superior**.

TO PULL YOURSELF UP BY YOUR BOOTSTRAPS also comes from earlier times, since a bootstrap was a loop at the back of the boot's neck through which a hook could be passed to help in the task of easing the boot over the heel. Although bootstraps have been around since the 16th Century, the use of the metaphor to describe how a man can **rise by his own efforts** was not recorded till the early 20th century.

By the way, an outmoded name for the thigh-high boots, once common, now only found on fly fishermen or thespians on the stage, was a 'buskin', giving us the rather nice term, 'Brother of the Buskin' to describe a fellow actor.

Mark Twain said with wisdom "A lie can make it halfway around the world before the truth has time to put its boots on."

BRACES ☞ BRASSIERE

An old French term, *bracière* probably came from the piece of protective armour worn around the arm – *bras* in French. This led to BRACE, which also means **to fasten tightly** or **shore up firmly**, hence to BRACE YOURSELF **to meet a challenge**. A BRACER is slang for **a nip of some strong spirit**, which might give you the temporary courage you need.

BRACES – suspenders to Americans – also **hold up men's trousers**, and there is an extra verse for the Davy Crocket song which goes:

> "Born on a satellite near the moon,
> His mother was a Martian, his father a Goon,
> He caught his braces on a passing rocket,
> And that was the end of Davy Crockett,
> King of the Universe."

BRACELET for **jewelled circlets for the arms** also comes from the French *bras*, hence the facetious term PUT THE BRACELETS ON HIM, CONSTABLE, indicating **handcuffs** the police might snap around your wrists.

BRASSIERE ☞ BRACES

Most of us call this garment a BRA for short; we borrowed that abbreviation from the French around 1910. In 17th century France, the *brassiere* was simply another word for a **bodice**, while *bracière* meant **a piece of armour for the** *bras* **or arm**. For us, this word denotes the "over-the-shoulder-boulder-holder" that lifts and separates a woman's breasts.

The development of the bra is a fascinating history in itself. To be brief, it arose (no pun intended) after long-overdue public health outcries about women's strangled internal organs, and miscarried pregnancies. The eventual creation of rubber-based elastic did away with constricting whale-boned busks and led to lighter, shorter corsets. This left the breasts needing additional girding-up, and thereafter, the modern type of brassiere evolved. Though the word came to us from France, the French word for this garment is a *soutien-gorge* or breast-support. Just think: without the invention of the bra as we know it, the whole idea of women's underwear could have been a flop!

Any feminist worth her salt might get upset at some of the modern children's rhymes recorded in Australian playgrounds, but they do have a childish exuberance and energy:

> "Ooh, Aah, off with the bra
> Down with the stockings and into the car."

BURNING YOUR BRA became a catchphrase in the 1970's **to define a Feminist**. The term originated in media hype of the time, and most people agree that no feminists actually burnt their bras until *after* the myth became current, when it was done as faked publicity. So the interesting question is: why did that gesture become so important? The reasons are complex. Firstly, it happened in the 1960s, when American men opposed to conscription to fight in the Vietnam War began illegally burning their draft cards and carrying out other (literally) inflammatory protests. It was also the era of The Pill when, for the first time, contraceptives allowed women sexual freedom without the fear of pregnancy, forced marriages, etc. There was a lot of discussion about greater equality in all aspects of life, and wearing, or not wearing, overtly gender-determined clothing. Most important, however, was that the BRA BURNING MYTH fed into the male breast fetish, and provided poor, threatened males with a way to trivialise the legitimate demands of the women's movement.

BRAT ☞ APRON

There is a Lancashire dialect word, *Brat, Bant* or *Brant*, meaning either an apron or the knitted cord to tie one. Since small, wilful children often had to be tied to mother's apron strings when she was working, I wondered if the exasperated, yet affectionate, term BRAT for a **tiresome child** might have come from this source? It is not mentioned as a possibility in the OED and other sources.

The usual suggestion is that BRAT as a word for a child may come from Irish dialect, *bratt*, for a **poor quality cloak** such as a beggar's child might wear. Either way, of course, your tiresome (but lovable) little BRATS would have a textile connection.

BREECHES or BRITCHES ☞ BROGUE and TROUSERS

In the same way as TROUSERS [qv], breeches is always plural. It probably derives from a word coined in in North Germanic prehistory speculated as *broks*, meaning leg covering; there are also links to the word 'break'. *Broks* evolved as breeches in England, but in Scotland, where the Viking influence was longer lived, people came to speak of putting on their *breeks* – and they still do.

Here is a colourful description of – I think – a trousered man in the rain, by one, Thomas Given, though I may be wrong, as only a Scot could understand it:

> "Was then he gaen his breeks a lift,
> then swung awa at east,
> while gingellin' hail an' wee spin drift
> tore through the nacke't trees."

Another dialect phrase is to describe someone as TOO BIG FOR THEIR BRITCHES. The spelling 'britches' came in by the 16th century, and the term probably came into use because someone bursting out of his trousers would look ridiculous, as we might also say swollen-headed. Possibly the first-known example of the phrase was written in 1835 by the American frontier hero, Davy Crockett.

Human skin has been referred to as 'hide' since the 17th century, and since tanning was the process of turning green skins into leather, TO TAN YOUR BRITCHES or YOUR HIDE offered a robust, yet marginally more polite way of saying, "I'll whip your buttocks."

There is a traditional English story about a happy gypsy named Boz'll and his dog; it's a nonsense tale full of absurd events. At the end, his dog dies, and his sorrowing master – never one to ignore an opportunity – mends a hole in the knee of his breeches with a piece of the animal's skin. "And that day twelvemonth, his breeches knee burst open and barked at him".

Another folktale tells us of a Viking hero called Ragnar, whose nickname, 'Loðbrok', meant 'Wooden-breeches'. The legend tells how he gained the name, courted Thora, a young woman who appears in some of the Norse Sagas, and spent seven years battling with Queen Gudrun. When Thora was a child, her father, King Heroth, found two little snakes in the forest and gave them to her as pets. Unfortunately, they grew into enormous, fire-breathing dragons who sizzled up any suitors who came calling on the girl. Thora fell in love with Ragnar and wove him a specially tough outfit of tunic and breeches, which she then left overnight in the river. When Ragnar came to fight the snakes next morning, the breeches were solid with ice, which meant that the creature's teeth and claws could not pierce them. What's more, when the dragons tried their best flame-throwing party trick on him, the ice melted and dowsed the fire, allowing Ragnar to chop the dragons down. It's such an elegant fairytale solution that I'm surprised it wasn't more widely used.

Right up to the 20th century, children's clothes tended to be smaller versions of adult attire, and in early times, they were often very uncomfortable and impractical for youngsters. There were marked differences for the sexes at about five years old. At that point, a boy would be BREECHED, which meant taken out of infant petticoats and **dressed in grown-up style breeches**. Often this coming of age rite would involve a ceremony in which the lad would be invested with a little sword and welcomed by the adult males. Little girls would get their first corsets at about the same age – poor things! To say a boy had been BREECHED became a shorthand way of saying that he **had joined the world of men**.

Children's clothes became simpler and more relaxed in the second half of the 18th century, following new ideas from Jean Jacques Rousseau on childrearing. Slowly, the term BREECHED for boys coming of age became obsolete. At around the same time, little girls began to wear soft, Indian muslins and were no longer strictly corseted. In lower-class families with working children, of course, fashion took a back seat to practicality. Just as childhood only gradually became a subject for scholarly study; interest in what children wore is something quite recent.

BROGUE ☞ BREECHES and TROUSERS

Oddly enough the word BROGUES, always used in the plural, came from the same source as britches – the Old Norse *brok*, but originally meant **rough shoes** that were worn in the remote Celtic areas of Scotland and Ireland.

The word first entered the English language in the late 16th century, but as a generic word for **a stout country shoe**, much later, in the 20th century. It seems likely that using the term BROGUE, meaning an **Irish accent**, travelled up from the feet to the tongue in the early 18th century – a typically Irish journey, you might say.

BUCKLE

The word BUCKLE originated as *buccula*, the little leather strap worn by ancient Roman soldiers, which passed across the cheek and held their helmets on. The muscles and other parts of the cheek are still referred to anatomically as 'buccal'. Fairly quickly, the term became associated with the metal ring and tongue used to fasten all sorts of things such as belts, possibly sliding into our language through the Normans. Old French *bocle* meant the raised, round boss of a shield, and we still talk about 'fighting with a sword and buckler', where the latter was a small, round shield.

Modern French *boucle* denotes any loop, but *bouclé* with an accent suggests curvy, circular objects – among many other things, curly hair, and both loopy yarn and the fabric made from it (so there's another textile connection). You might save this phrase for your next holiday in France: "*La boucle est bouclée*" translates as, "We've come full circle" – probably the result of a male driver who refuses to ask for directions, even though he's lost.

It must be the word connection with shields and armour – and once again, the French verb for something that bulges – *bocler* – that gave us TO BUCKLE, meaning **to bend** or **give way**.

Saying you should BUCKLE DOWN TO A TASK could mean you feel you need to gird yourself in full Roman armour before doing the washing up or a little light dusting. Or it may just be a variant of 'knuckle down' meaning **to address yourself to the task**.

The humble BUCKLE as a device for fastening has an important place in rhymes too, such as the counting ONE TWO, BUCKLE MY SHOE, and the traditional:

> "Bobby Shafto's gone to sea,
> Silver buckles on his knee,
> He'll come back and marry me,
> Bonnie Bobby Shafto."

BUDGET ☞ PURSE, POCKET, BAG

When we try to keep our finances within set limits, in other words, set a BUDGET, it should be realised that we are really keeping our spending within a metaphorical purse, sack, leather bag or even womb, since the word comes from Latin *bulga*, meaning all these things. BUDGET probably came to us in the 15th century, from the French word *bouge*, with a diminutive of *bougette*. Alas, for many of us, one only needs a diminutive bag to hold one's worldly assets.

A slightly old-fashioned phrase that can still be heard sometimes is to OPEN MY BUDGET, meaning **I am going to speak my mind**. I'd like to think that wisdom is priceless, but I doubt if it's gold that tumbles out in most cases when I do this.

{*This word can also be found in* **Rigmaroles and Ragamuffins**, *see entry under* EXCHEQUER}

BUREAUCRACY

We tend to have a very negative image of BUREAUCRATS, although really it all depends on how well this collection of officials who comprise a BUREAUCRACY work together. If they are efficient, co-operative and helpful to the public, life can be sweet. Alright – so it isn't always like that, but what on earth have they to do with textiles?

A BUREAU, both in France, the source of this word, and from the mid-18th century onward, also in the UK, was a writing desk, usually with drawers in which to keep papers. The name derives from *burel* in Old French a coarse woollen fabric, frequently dark brown, used to cover the writing surface. This sopped up ink stains and prevented the pressure of pens damaging the wood. In England, such fabric was known as 'baize'. It was a soft, loosely woven cloth, often dyed green or maroon, and sometimes felted.

An alternative explanation is that the word *baize* came from Baza, Spain, where it was originally made. Perhaps. However, the French version was known as *Bure*, and this name became synonymous with a more specialised – perhaps even a superior – product.

Either way, the word BURE shifted from describing the covering **to the desk itself**, hence the BUREAU, and by one easy step to **the man who sat and worked at the desk**, a BUREAUCRAT. **A collection** of BUREAUCRATS constitutes a BUREAUCRACY, and these days, they are often – maddeningly – far away from their desks when we try to telephone or email them.

It is probable that the word BURGEONING to describe plants and shrubs proliferating and bursting into flower may also derive from a textile origin, and again, from France. There's a late Latin word, *burra*, very similar to *burel*, that meant 'a tuft of wool'. Eventually, this gave rise to the old French verb *bourgeonner*. The word *burjon* sprang from this, meaning **to produce buds**, and so on finally to the idea of **plants sprouting and becoming lush**. We use the term TO BURGEON figuratively about all kinds of things that flourish: **ideas, hopes, careers,** or any **new beginnings**.

CAMIKNICKERS ☞ KNICKERS

Camiknickers belong to the vanished world of mildly-saucy novels from the 1920's and 30's, when gorgeous girls reclined on satin coverlets in glamorous boudoirs. And who has either of those now? Mind you, I'm sure there are still many who secretly own camiknickers – more commonly known these days as a 'teddy'. (Come on, own up; you know who you are!) But for anyone who hasn't seen such things, they were (and are) one-piece silky undergarments of vest and knickers combined, often with lace insertions and a frilly trim. The name was first seen in print in 1915.

So, where did the idea spring from? Well, armoured soldiers in 16th century Spain had a problem: how to recognise friend from foe when everyone was covered from head to toe in bits of metal. They solved it by donning shirts with highly-visible symbols and colours over their armour, and adapted the Arabic word *kamis*, meaning an undertunic, as the Spanish *camisa* or shirt. To fight *camisada* literally meant **to attack in one's night shirt**, because they used guerrilla tactics and ambushed enemies at night.

For the same reason, in 18th century France, Protestant Huguenot rebels became known as the *Camisards*, which simply meant, in their local Occitan tongue, that they fought in peasant smocks. Their *camisas* evolved into the modern French *chemise*, denoting **a loose-fitting dress like a slip** that falls straight from the shoulder, or **an undergarment for a lady's top half**.

All things French – and lingerie in particular – tend to be regarded as sexy. Yet, perversely, the British reclaimed the *chemise* as a CAMISOLE; when the lacy top and drawers were combined as a single garment, it seemed logical to call the result CAMIKNICKERS.

Incidentally, from the 18th century on, soldiers coined a slangy word to refer to their **battle shirts** as A COMMISSION or a MISH, also from the *camisa* although this term gradually fell out of use in the early 20th century.

What a strange path for a garment to take over four hundred years – from a bloodstained shirt for a sweaty Spanish soldier, to a wispy undergarment on a 20th century lovely and back to a warrior's shirt again!

CANOPY ☞ CANVAS, PAVILION, SAIL and TENT

The word CANOPY for a tented or projecting cover sounds as if it should link with *cannapacem*, Latin for a sail – and maybe it does, somewhere along the line. However, its direct origin lies in the Greek *konops*. This meant a mosquito, and eventually evolved into *canopeum* in medieval Latin, denoting both a **couch** and the **mosquito net covering it**. What a nifty bit of garden furniture that would be, now that global warming is filling the great outdoors with clouds of new and annoying insects. Please, designers, hurry up and bring this one back!

There are similar words which probably relate to others from the mists of time, which sound similar. *Kanab* in Persian and *xanapiz* in Greek refer to the plant that we know as hemp and which provides cannabis. Indeed, hemp is also linked with CANVAS and may well have been used to make the covering.

CANOPY also then stretched onward to mean **any covering extended over a throne or couch, to show off the importance** of the occupant. The use of fabric for this purpose, as well as shade and protection from insects, was known in ancient Egypt, Greece and the Roman world, and the word still tends to carry a sense of posh pageantry or formality.

In French, however, the original bug-screened couch shrank and became CANAPÉ, **a small piece of bread** on which a delicious **topping** reclines at its leisure. If you attend the kind of drinks parties that serve these twee refreshments instead of proper food, perhaps you could imagine yourself eating mosquito on toast so as to keep the original word alive. Or Uuuugh.... perhaps not!

We also use the term CANOPY metaphorically for any sort of **overarching covering**, as in THE CANOPY OF THE TREES and CANOPIED WITH SUNSHINE; such phrases can sound poetic or slightly pretentious depending on your point of view.

CANVAS ☞ CANOPY, PAVILION, SAIL and TENT

The rough-woven, unbleached cloth we call CANVAS probably originated in Persia or Greece, as it was once made from woven hemp found in both those countries and their former empires. *Kanab* in Persian and *xanapiz* in Greek for what they called the plant – see under CANOPY.

{See *also my first book,* **Rigmaroles and Ragamuffins**}

The same useful fabric is used to make tents and sails for ships. To say someone is UNDER CANVAS suggests they are **camping out in a tent**, but to say a ship is UNDER CANVAS means its **sails are up** and it is ready to run before the wind.

Because a human body can also be supported by a large sheet of canvas the verb TO CANVAS can be used in the sense of **tossing someone about roughly in a canvas sheet**, to **buffet t**hem, even to beat or batter them. It can be used metaphorically for **criticising someone's written words**, or **to start up a critical argument with them**, extended to negotiating, and persuading.

It is presumably in the latter sense that we use it to talk of politicians **canvassing** for votes; I'm not sure they'd get many if they tossed their electorate around in a rough and violent manner. They save that to do figuratively once they've been elected, I guess.

CAP ☞ CAPE, CAPPUCCINO, CHAPEL, CASSOCK, CLOAK, COAT and MANTLE

The word CAP is almost certainly derived from the Latin *caput*, for the head, and has given rise to many related words or phrases, far beyond the literal meaning of a covering for a person's head. Almost anything that is used **to cover, complete** or **fit on the end** of something else can be described as A CAP.

You might use it metaphorically, as in, TO PUT A CAP ON EXPENDITURE or IMMIGRATION. Or even TO PUT A CAP ON THE NUMBER OF MEANINGLESS SPEECHES WITH WHICH LONG-WINDED POLITICIANS ARE ALLOWED TO BORE US.

Used as a verb, TO CAP a story, quotation, joke or verse, is **to come out with another** even better example or punchline. THAT CAPS IT! is another way of saying, **that completes something or surpasses** whatever came before with something **excellent**.

To PUT ON YOUR THINKING CAP means, as it sounds, **to give a problem your full attention and think it through carefully**. The original phrase seems to have been CONSIDERING CAP, only being changed in the 19th century. It is described in the fictional 'The History of Little Goody Two-Shoes,' of 1765, '... a considering Cap, almost as large as a Grenadier's, but of three equal Sides; on the first of which was written, I MAY BE WRONG; on the second, IT IS FIFTY TO ONE BUT YOU ARE; and on the third, I'LL CONSIDER OF IT.' That sounds to me a useful, if rather sanctimonious idea.

In my childhood, we used to play a game at Christmas which started with everyone being given a colour for a name – Mr Green, Miss Blue, Mrs Red, and so on. Then, one person would start by saying "The Priest of the Parish has lost his Considering Cap. Some say this and some say that, but I say Mr Green [or Miss Blue, or Mrs Red] has it." The person named then started a little dialogue, "What I, Sir?" "Yes you, Sir" "Not I, Sir!" "Who then, Sir?" "I say so-and-so" naming another person, who went through the same process. When I was small, I loved the almost poetic, antique cadence of the words, even as I stumbled over them. The grown-ups of course were better at it – until the Christmas wine and brandy took their inevitable toll and inattentive cousins and ancient great uncles one by one fumbled the phrases and were declared "out!"

The CAP AND BELLS were the insignia of the Fool or jester, so TO PUT ON A FOOL'S CAP is **to play the fool**. Incidentally, the paper in the size known as "foolscap" has no textile connections. Its name derives from an old watermark showing a jester's jingly headgear.

An old fashioned way of saying that a woman had decided **to tempt a man into matrimony** was to say SHE SET HER CAP AT HIM. I think it's overly optimistic to think that all that is needed is a pretty hat at a jaunty angle; surely men aren't such fools as that? Or are they?

In Jane Austen's day, every young woman was encouraged to catch an eligible husband (and if possible, a fortune) before she lost her looks and vivacity. Her 1811 book, *Sense and Sensibility*, contains this passage:

> "Aye, aye, I see how it will be," said Sir John, "I see how it will be. You will be setting your cap at him [Willoughby] now, and never think of poor Brandon."

"That is an expression, Sir John," said Marianne, warmly, "which I particularly dislike. I abhor every common-place phrase by which wit is intended; and 'setting one's cap at a man,' or 'making a conquest,' are the most odious of all."

IF THE CAP FITS, WEAR IT, is a gentle way of telling someone offended by a comment that **it's probably true**. TO CAST, or THROW YOUR CAP INTO THE WIND is to give up or let go – perhaps **surrendering to despair**, or **taking chances** by abandoning conventional behaviour.

CAPARISON ☞ CAPE and CLOAK

This rich-sounding word shares its origin with cape and cloak. A *caparasso* was a hooded cloak in 16th century Provence; but in other times and places, it could be an old woman's cloak or a covering for a pack-horse saddle. If highly decorated, it suggested a measure of wealth. We might say, in a flowery, poetic moment, for instance, while writing a historical novel, that the protagonist was CAPARISONED IN BEAUTIFUL ROBES. However, if your heroine was well read in word origins, she might not be entirely happy to be likened to **a horse decked out in useless bits and bobs**.

Why? Because the word is closely linked to **trappings**, meaning the extra ornaments on the harness of a horse – such things as bells and brasses, non-essential finery. The word harness is itself linked to **draper**, **drapes** and **drab**.

{*For more details about these words see my first book,* **Rigmaroles and Ragamuffins** *for the entry under* DRAB}

CAPE ☞ CAP, CAPPUCCINO, CHAPEL, CASSOCK, CLOAK, COAT and MANTLE

The word for a cape in its most common meaning today of a cloak – with or without a hood, probably comes from *caput*, Latin for head and relates to cap. However, possibly it is more closely related to *capere*, meaning to take, and links to words meaning able to take, hold onto, or have room for – such as capacious, capable, and capacity.

A CAPE'S other common significance is **any headland or promontory jutting into the sea**. Some well-known examples, such as the Cape of Good Hope in South Africa, give their name to **a whole region**, known simply as THE CAPE.

A Cape or a Cloak may be the most common names for an outer, wrap-around garment without sleeves, but we have quite a lot of other similar words – not surprisingly, since it is such an important article of clothing.

Quite dashingly, the peace-loving clergyman's CASSOCK is etymologically really a bellicose Cossack thanks to the Russian root, *kozak,* and only became wrapped round the pious churchman in the 17th century. The word can be traced back even further either to Turkish *quzzak,* meaning **a vagabond** or **nomad**, or to Persian *kahaghand,* which was **a padded silk jacket**.

In Italian the Latin word for cloak, *cappa,* developed into *scappare* which probably gave us the slang word, SCARPER, also meaning **to get away, to leg it out of trouble, to vamoose**. In rhyming slang, 'Scapa Flow' was used to mean 'go' from the 19th century on but was reinforced when the German fleet was scuttled there there at the end of the First World War and the word probably became linked with the earlier SCARPER.

Words such as capricious and caper probably derive from capriole, which relates to goats and the way they prance, rather than to caput, the head. Sadly, therefore, they have no textile origin.

CAPPUCCINO ☞ CAPE, CHAPEL, CASSOCK, CLOAK, COAT, GRAY EMINENCE and MANTLE

CAPPUCCINO comes from an Italian word for a hooded cloak, and is the diminutive form meaning 'a small hood'. In 1528, a new order of Franciscan Friars was instituted in Italy, becoming known as *Capuchins* from the distinctive pointed hood of their habits. It took 200 years for women to appropriate cloaks with a similar hood as fashion garments in the mid-18th Century.

From there, I'm afraid it was downhill all the way, until the name was given to a novel drink containing **espresso coffee, hot milk and steamed, milky froth** which is now served all over the world. Its creamy-brown colour resembled that of the Friars' habits, and the foam on the coffee also stood up in a peak, reminiscent of the pointy hood.

European explorers venturing into the wilds of Central America in the 15th century, first applied the word capuchin to a **monkey** with a pointy tuft of black hair, and it has also been attached to a breed of **pigeon**. A **capuchin's beard** is neither fur nor fowl, but a slightly fluffy form of salad endive with toothed leaves.

If you wanted to pay homage to the French Franciscans, I suppose you could sit in a cafe wearing a **cloak**, with a **monkey** on your lap, a **pigeon** on your head, eating **endive** salad and drinking a delicious CAPPUCCINO. No doubt, the monks would appreciate the joke, but your cafe owner might ask you politely to move on.

CARPET ☞ RUG

The word that has become our CARPET, denoting almost any form of floor covering, probably originated in ancient Armenia. Perhaps that isn't too surprising, since its native folk are among the greatest craftspeople and artists in the world, carrying their skills everywhere in their worldwide diaspora. They had two relevant words, in fact: *karpet,* which usually described a smooth, non-pile covering, and *gorg,* for tufted rugs.

Kap for a knot, which is the most common technique in making patterned carpets and rugs, probably came from a dialect word in 5th century Armenia. As I am also a Kapp, I rather like to imagine my ancestors were knotty little Armenian carpet makers, though it's equally possible they were cape makers. In medieval Latin, a *carpita* was a woollen counterpane, and it is probable that the word came into usage in both Britain and France some time in the 13th Century, a momentous era for trade with the East. In any case, carpets are what we decided to put on our floors after we realised that rushes, bits of chewed-up animal, and excrement did not make for civilised living.

An alternative source for the word suggests that it derives from *carpere,* Latin for to pluck, referring to a cloth that had been plucked, divided or shredded, the same word as gave us HARVEST. People are always trying to be clever by using Latin phrases like *carpe diem*. Actually, what Horace wrote in his "Odes", in 65 BC was: *Carpe diem quam minimum credula postero* or "Pluck the day and place no trust in tomorrow." Certainly, grasping a nice, juicy piece of fruit while it's dangling before your nose makes perfect sense – as long as you don't dribble the juice on the carpet!

Whatever the source, CARPETS – especially richly figured Ottoman pieces – were originally table coverings, and not cast onto floors until the 15th Century, when one sees them under the feet of saints, nobles and the Virgin Mary in a host of Renaissance paintings.

Then, gradually the carpet began to move entirely underfoot, although it remained a symbol of prestige and affluence. The word eventually became a figure of speech for **anything resembling a carpet in smoothness, softness or colouring**, such as A CARPET OF AUTUMN LEAVES or a meadow **rich in wild flowers** described as CARPETED WITH FLOWERS. It recalls a charming Arabic saying with the same meaning, WIDE IS THE CARPET OF SUMMER.

A KNIGHT OF THE CARPET was a man who was **made a knight in peacetime**, not dubbed on the field of battle. The term A CARPET KNIGHT was a contemptuous 16th century insult aimed at **stay-at-home soldiers**; they probably looked very decorative in their uniforms and armour, but wouldn't want to get mussed up by actually fighting.

A CARPETBAGGER was another derogatory term, and came from the type of **cheap travelling holdall** made of fabric and called a CARPET BAG which was carried by opportunistic Americans from the North who went South after the end of the Civil War in 1865, seeking to profit from negro emancipation both financially and politically. Hence, a politician **on the make**, or one who **seeks office where he has no community connections** is still accused of CARPETBAGGING.

You might say to someone, "I'LL HAVE YOU ON THE CARPET!" which contains a double entendre that makes it a good addition to an otherwise literally downtrodden subject. Either someone is about to **tell you off big-time for some wrong**, or **to have their way with you sexually**. Please don't start slipping out of your clothes until you're quite sure.

When we speak of TO CARPET someone or have him ON THE CARPET in the sense of **giving him a good dressing down** (see DRESS) it seems usually to occur in the context of work. It's a phrase of fairly recent origin. One suggestion derives from the shift of the carpet from the table to the floor. It could be that, in the early 20th century, a servant or an employee would be summoned to the only bit of the room that was prestigious enough to possess carpeting, i.e. near the hearth in the drawing room or the patch of floor around the boss's desk.

If those explanations don't sound very convincing, how about another version, which says that the term TO BE CARPETED came from the highly status-conscious civil service; it meant to be moved down from **a floor with a carpet** to **one with bare boards**, clearly a demotion. Somehow, CARPETED sounds more like a promotion to having a carpet. Oh well. Tuft luck.

If you are asked if you would like to CUT A CARPET or CUT A RUG, in the USA, please don't get down on the floor with your scissors, to the horror of your hostess. It is 20th century slang for **dancing energetically**.

We talk about ROLLING OUT THE RED CARPET on **great occasions to greet honoured guests**, as special decorations under foot have been used for hundreds of years. The red carpet for civic ceremonies and its use as a figure of speech have been noted since at least the 1930's.

Not all carpet-related phrases have happy associations. CARPET BOMBING is the practice of bombing an area from planes to such a degree that **everything in an area is destroyed**. To say THERE MAY BE BLOOD ON THE CARPET suggests a meeting in which – to use another metaphor altogether – the knives are out and **someone's going to get hurt** – or more probably, SACKED [qv under SACK].

Let's find a MAGIC CARPET and fly away from all our troubles, like heroes and heroines in the Arabian Nights and other Fairy Tales.

CHAPEL ☞ CAPE, CAPPUCCINO, CASSOCK, CHAPERONE, COAT, ESCAPE and MANTLE

We have seen how the *cappa* from the Latin *caput*, meaning head, was a cloak with a nice warm hood to keep said head cosy. In the 18th century, we find the word COPE – as in a bishop's semi-circular cape. From CHAPEL we also get CAPPELLA – which spins off into a style of singing – *a cappella* – meaning **without music accompaniment**, and recalls centuries of Gregorian chant or Plainsong, when the human voice was the only 'musical instrument' used in the chapel.

In fact, the word CHAPEL has an even more direct textile link, and it was born at Tours, France. In the 4th century AD, a young soldier named Martin is said to have been so touched by the plight of a shivering beggar that he cut his expensive *cappella* – his military cloak – in two with his sword and gave half to the man. Subsequently, Martin became a monk and, in due course, was sanctified. At Tours, the guardian of the precious relic of the saint's putative half cloak was called the *capellanus*, latinised from Old French *chapele*.

This offshoot also gave us CHAPLAIN, meaning **a clergyman attached to a private chapel, ship, military regiment** etc. Probably also related is CHAPLET, meaning either **a crown of roses** or **a flowery circlet** for the head or, to bring things neatly back to religion, **a string of 55 rosary beads** used for counting prayers.

CHAPERONE ☞ CAPE, CAPPUCINO, CASSOCK, COAT, ESCAPE and MANTLE

Here's another descendent of that prolific Latin parent, *cappa*. When young couples were courting in the 18th century, CHAPERONE arose to denote the **protective presence of a mature female attendant**, who made sure nothing too naughty happened. Such women, being either married or widowed, concealed their hair modestly under some form of HOOD or CAP, whereas unmarried girls could leave their hair loose or uncovered.

Actually, the earlier French form, *CHAPERON*, meant **a round headdress of stuffed cloth with wide cloth streamers** that fall from the crown or are draped around it, worn in the 15th Century. Were 18th century youngsters trying to imply their elderly aunt was a 'fathead'?

CHRISTMAS STOCKING

Although this phrase is not often or necessarily used on its own, the words CHRISTMAS STOCKING instantly conjure enchantment and delight for almost any child in the Christianised Western world. Traditionally, it is **a stocking, pillowcase or bag** hung from the mantelpiece or at the foot of a little one's bed on Christmas Eve for Santa Claus, St Nicholas or The Three Kings (depending on the country) to fill mysteriously in the night. Next morning, the container spills out its lovely contents – little gifts, sweets, and a deliciously-scented orange or tangerine right in the toe.

Santa Claus derives from Saint Nicholas, a 3rd century saint born in Patara, which was then Greece, but is now part of Turkey. As quite a young man, he became Bishop of Myra, and many wonderful stories were told about him down the centuries. He was renowned for his love of children and his generosity. To save three daughters who had no marriage dowry from being sold into slavery, he chucked golden coins or bags of coins through their window, which landed in shoes left to dry before the fire.

The historical St Nicholas is regarded with reverence and affection by most Christians, and he is the patron saint of an eclectic group including children, sailors, fishermen, merchants, archers, thieves, and students. He also protects pawnbrokers, whose traditional sign of three gold balls hanging above the door commemorates the most famous story about the saint.

Somewhere along the line, St Nicholas became conflated with a benevolent Norse nature figure, Father Christmas, who makes everyone happy and jolly in the dark days of winter. A Coca Cola advertisement changed his traditional long, green coat to fur trimmed red one, worn with a bobble hat!

CINCH ☞ GIRDLE

This term seems to have been coined in the mid 19th Century, probably from the Mexican word for a saddle girth, since it is a band, cord or belt that can be fastened tightly and securely.

It gained currency in the USA as IT'S A CINCH! meaning **it's easily done**, and hence a phrase suggesting **a sure thing**. This link to the Wild West goes down a treat in the UK too. "IT'S A CINCH WE'LL GET THIS DEAL!" says a businessman, narrowing his eyes and imitating the rolling gait of the Frontiersman. Such antics always look very silly in an English pinstripe suit, but don't tell the poor crittur that; gol dern it, let him dream.

CLOAK ☞ CAPE

The traditional CLOAK, worn by men and women from ancient times to the present day, can be made of any material, but is usually fairly thick and strong. It is sleeveless, maybe having slits in the sides for the arms, and may fasten down the front, or just at the neck or shoulder, allowing it to be thrown back to free the hands. Whether long or short, a CLOAK is generally a loose outdoor garment covering other clothing – hence its link with disguise, subterfuge and cover-ups.

You might CLOAK your **embarrassment** or **fear** or **hatred**, or indeed almost any other **emotion, intention** or **Cunning Plan**. We can say the countryside is CLOAKED IN DARKNESS, among many other metaphorical uses.

It is advisable to CUT YOUR COAT ACCORDING TO YOUR CLOTH, which of course means **to work out if you have enough fabric before cutting it out**, in case there isn't enough. It tends to mean, colloquially something more like "Don't be extravagant" or "Think things through before setting out on a project".

What about CLOAK-AND-DAGGER-WORK? Aha! This mean **spying** – James Bond stuff. It comes from a 16th century phrase in Spanish and refers to exciting plays of the day full of fights and intrigue, in which actors often concealed weapons beneath their cloaks. These stirring entertainments were popular in England too, until the Puritans banned them in 1642. Even though cloaks have more or less disappeared, spies still go 'undercover' with dark intentions, hidden identities and ever more sinister weapons.

Another ominous link with CLOAK is the bacterium Chlamydia, varieties of which cause infections such as psittacosis in birds, and genital infection in humans. The bacillus, when first seen under the microscope, was thought to CLOAK **the nucleus of the host's cells** – erroneously, as it happens – and was therefore named CHLAMYDIA after the *chlamys* or CLOAK worn by hunters and military men in Ancient Greece.

We discover a much more up-to-date connection in a CLOAKING DEVICE. This is the stuff of fantasy: a science fiction **device that can render any person or object totally invisibl**e. Many stories, from old fairy tales to modern Harry Potter have used defensive CLOAKS OF INVISIBILITY. Some imaginative authors have made buildings, spaceships, planes and whole towns disappear. Before you guffaw too loudly, scientists have recently managed to bend at least one wavelength of light around objects in the laboratory, and this so-called Active Camouflage may become reality in the near future!

CLOGS ☞ SABOTS

The origin of this word is unknown, although *clogge* to mean a lump of wood appears in English as long ago as the 14th century, and may have Norse roots. Apparently, it was also rude medieval slang for a testicle.

CLOGS are traditional wooden pattens or shoes with wooden soles. They're usually rather coarse and very utilitarian – in other words, peasant wear, mainly for mucky outdoor work. Such CLOGS feature in a number of current British sayings, such as TO POP YOUR CLOGS, which means **to die**. To pop something can also mean **to pawn it**, as in "pop goes the weasel" where "weasel and stoat" was Cockney rhyming slang for a coat.

If someone wants to put you in your place gently, they might say you are **"too clever for your own good,"** or **"too clever by half"** – or simply call you a CLEVER CLOGS.

In Lancashire, where working folk such as millworkers were once habitually shod in hard-wearing clogs, the saying we noted earlier about rags to riches in three generations emerged as "CLOGS TO CLOGS IS ONLY THREE GENERATIONS."

Of course, TO CLOG also means **to impede or encumber something**, as in CLOGGING UP THE SINK with those nasty bits of potato or hair in the plug hole. And TO CLOG UP THE WORKS denotes something (or someone) that prevents machinery or any normal process from working properly.

CLOTHES

We call body coverings CLOTHING. Human beings probably started to clothe themselves by gathering materials such as leaves and animal skins and joining them together using plant or animal fibres, just as Adam and Eve were said to have done. You might assume that this was obviously to keep warm in cold climates, and protect against the sun in hot ones. Not entirely true. Anthropologists generally agree that from the earliest known times – and even in the most agreeable climates – we humans have used both body decoration and clothing to show off our wealth, signal our availability (or previous commitment).

CLOTHES have always expressed our creativity and individuality. Inanimate objects can speak louder than words – especially when making a first impression, so our ancestors continuously flaunted their dress, jewelry, body paintings, tattoos and so on. Even a visitor from far away could understand their messages: "I'm the boss", or "I'm rich", or "I'm married to that ape-man over there, so watch out". Personal embellishments provided a whole

··· *Clothes* ···

alternative language – a universal and eloquent way of communicating that we still use today.

As you would expect, the English language is full of colourful sayings and figures of speech relating people to their clothing. That's because CLOTHING is the most intimate of the many connections that we make with thread, and CLOTHES form an important and emotive part of the lives of even the least fashion-conscious individual.

Some authorities link the word CLOTH to an ancient Goddess, the first of the Three Fates, who spins the thread of our lives. She was *Clotho*, a Greek word meaning I twist. The Danish *klud* for a rag and the german *kleid* meaning a woman's dress are linked to the Irish, *cludach*, for a covering – k and c are often interchangeable in European languages.

{*For more about* CLOTH *see my first book,* **Rigmaroles and Ragamuffins**}

One can BE CLOTHED WITH a lot of things figuratively, such as **shame**, **righteousness**, **beauty**, **goodness**, and many more. Religious folk who consider they are 'saved' may be metaphorically CLOTHED in **glory**, or **light**.

We use the term A CLOTHES HORSE not just for the wooden-slatted contraption on which clothes are aired, but also to describe **a vain woman** who seems to exist only to show off her attire and is of no importance otherwise.

Talking about THE CLOTH when you mean **the clergy** has a nice old-fashioned ring. Sometimes we speak about A MAN OF THE CLOTH, a phrase that conjures up snug country parsonages and the sort of solid, respectable **vicars** who feature in Agatha Christie whodunits.

Dozy old CLOTH EARS means someone who is **not listening to** you or is **stupid**, and the related CLOTHEAD is **a fool**.

An interesting feature of folklore in several cultures is the idea of CLOTHING GIVEN TO THE BROWNIES – a type of house fairy or supernatural helper. Legendary creatures from other realms often seem to be nude or dressed in animal skin, bark or other natural stuff. This seems to link with ideas of them as primitive or pagan creatures. Some fairy folk are given garments and then dance away, either offended or joyful, never to be seen again. In other folk traditions, clothing transforms them into a more 'civilised' state or makes them more 'human' – which could be a blessing or a curse, hence the ambivalence of many stories.

"Tell her to make me a cambric shirt without any stitch nor needlework…" sings a man about his true love, in the folk song 'Scarborough Fair'. Is this a message about supernatural powers, as well as setting the girl a playfully impossible task? In some versions, she replies with equally impossible tasks. "Tell him to find me an acre of land/ Betwixt the salt water and the sea strand."

This is a very old ballad with many versions, sung in both England and Scotland. The siting of it at Scarborough Fair and the refrain, 'Parsley Sage, Rosemary and Thyme' were added in the 19th century. Now, perhaps, the song is best known through the Simon and Garfunkel recording, but it has also been sung by many others. Both the minor-key tune and its words are particularly haunting.

CLOTHING is a key feature in so many myths and fairy tales. One story has its protagonist dressed only in a donkey skin throughout her childhood. In another, when struck with a glove, the fairy lady of Llyn y Fan Fach must return to her Welsh lake. The Seven Swans, actually enchanted brothers, recover their human form when the heroine throws a shirt she has made from the fibrous, linen-like stems of nettles over each of them. She has not had quite enough time to complete the final sleeve, so the youngest boy bears one swan's wing instead of an arm ever after.

One of Aesop's best known fables tells about a hungry wolf. The shepherd and his dog were too watchful to allow him near the flock. So the wolf bound a sheepskin onto his back and gradually crept nearer and nearer. The Shepherd, lulled by appearances, didn't notice until it was too late. Throwing off his disguise, the wolf sprang up with a howl, carrying away a nice, fat sheep. We still use the metaphor of A WOLF IN SHEEP'S CLOTHING for anyone who is both **deceitful and malevolent.**

Similarly, in the Bible [The Gospel according to Saint Matthew Chapter 7, verse 15], Jesus warns his followers, "Beware of the false prophets, who come to you in sheep's clothing, but inwardly they are ravening wolves."

CLOTHES MAKE THE MAN

We sometimes describe a person by what they wear, employing a word or phrase that supposedly sums up the person and his lifestyle, and eventually becomes common parlance. Usually, this form of shorthand slang is derogatory, and may have class overtones.

"THE GARMENT MAKES THE MAN" was a catchphrase in ancient Greece, and is recorded in England from the 16th century on. Shakespeare nabbed this concept for his play Hamlet, in which Polonius advises his son to dress in a conservative, yet impressive and 'socially accepted' manner:

"Costly thy habit as thy purse can buy,
But not expressed in fancy—rich, not gaudy,
For the apparel oft proclaims the man,
And they in France of the best rank and station
Are of a most select and generous chief in that."

Later, the idea was sometimes stated as "THE TAYLOR MAKES THE MAN", but the implication remained the same: clothes not only determine how strangers see us, but can actually help us to 'grow into' the image they convey.

CLOUTS

This word may have come originally from the Dutch *kluit* which meant a lump and became in Old English a *clūt,* for a patch in cloth, or even a metal plate. It is therefore related to later words such as lump, clod, cleat and clot.

At one time, CLOUTS was a slang word for women's below-the-waist underwear, as well as for a handkerchief, a sanitary pad, or even **a thief specialising in hankies** – a CLOUTER. Also, of course, A CLOUT could mean **a short, sharp blow** – as used colloquially since the 19th Century.

It was even a term, CLOUTED CREAM for what we now call **clotted cream**, perhaps because to make that lovely, thick, tasty treat, the cream skimmed from the top of the milk must be heated slowly by indirect heat and then cooled very gently, as well till the clots rise.

The traditional Scottish CLOOTIE DUMPLING, a **fruity suet pudding**, is tied up in a **little cloth** or *clootie* and steamed in a pot. It could make for difficulties if one doesn't watch one's pronunciation. I would be most upset if I asked for a clootie dumpling – or for clotted cream, come to that – and received instead a sharp blow to the head!

CAST NOT A CLOUT TILL MAY BE OUT is a well known English proverb – though there are versions in French and Spanish too. The meaning is quite clear – **don't leave off your winter woollies** [especially the vests and long johns] **until the weather allows it safely**. However, there is always bickering about whether 'May' means the month, or the old term for hawthorn blossoms, as in the name of the Pilgrim ship, Mayflower. Since the hawthorn blooms round about April, you can choose when to strip off. Just be careful how close you get to the tree: hawthorns have vicious prickles that harbour pathogenic bacteria which can give some people nasty symptoms!

Oh bother it! I'll just keep my big, warm Bridget Jones bloomers on all year round.

COAT ☞ ARMOUR, CLOAK, JACKET, JUMPER and PETTICOAT

A COAT today is generally understood to be an outer garment which covers the upper half of the body, but may be any length, to the calf, ankles or feet. It has sleeves for the arms, usually full length, as it is primarily a practical garment for protection against the weather. Ancient Persians were probably the first to introduce garments with sleeves, seams and front fastenings to keep the warmth in. Sadly, we don't know what they called this useful garb.

It can be difficult to make sense of the changes in words describing pieces of clothing through the ages, but the word we use for warm outerwear – A COAT, has been traced back to the Roman *cotta,* a man's over garment; this was handed down to us (no doubt very frayed by then) via Old French *cote.* This and its related word *cotte,* which meant 'a rib', had always suggested something worn close to the chest. And, early on, true enough, *cote* described a sort of short, close-fitting cloth tunic with sleeves worn as a male undergarment.

However, in the 16th century, its lower part was referred to as the *jup,* a French adoption from Arabic *jubbah,* a flowing robe. Perhaps that is how humble underwear moved up in the world, and a full *cote* grew into a longish outer coat for men. We have kept something of this idea by speaking of any coat with a lot of fabric in the lower part as having skirts. You see, the French – always a thrifty people – didn't want to throw out a perfectly good word, so they retained **jupe** for a woman's skirt, (see also JUMPER) but the word 'coat' denoting a skirt for women is now only preserved in PETTICOAT [qv].

In the early 19th century, coats began to be divided into OVERCOATS and UNDERCOATS. While we still use the former, the latter is rarely heard, except as an UNDERCOAT OF PAINT, which means of course the use of **a layer of other paint** put on first. Even so, we still give a nod to this idea in garments such as a HOUSECOAT, or in the phrase TO WEAR A COAT AND TIE, or TIE AND TAILS – no longer a heavy coat for warmth, but a snazzy **black Dinner Jacket** for gentlemen. (see JACKET)

We do not need to concern ourselves with such sartorial exotica as lounge coats, morning coats, dress coats, tailcoats, sports coats and frock coats, but it's worth mentioning the phrase describing someone CLIMBING OR RIDING ON THE COATTAILS of someone else. It means that they may be guilty of **stealing someone else's ideas** or **benefiting from another's hard work** rather than making a success of things through their own efforts or talent. It's a graphic image not in the least diminished by today's paucity of coats with tails, except at weddings and funerals.

Because a coat is something that covers part or most of the body we use TO COAT, or COATING as a figure of speech for almost anything that **covers or conceals**. We can say that THE LANDSCAPE WAS COATED WITH SNOW, or describe an onion or other vegetable as HAVING AN OUTER COAT. Almost anything can be COATED, whether it's **a wall with paint**, an **animal with fur**, or **a slug with slime**. Even someone's tongue can be **coated** with a horrible feeling due to a hangover. Not that this applies to you or me, who never overindulge.

If you tell someone he should CUT HIS COAT ACCORDING TO HIS CLOTH you are suggesting **he should try to live within his means financially**, or generally **make the best of what he has**.

There is a very ancient Christmas carol about lending baby Jesus a fur coat:

> "Baby Jesus, sweetly sleep,
> Do not stir,
> We will lend a coat of fur;
> We will rock you, rock you, rock you,
> We will rock you, rock you, rock you,
> We will serve you all we can,
> Darling, darling little man."

And in lighter vein, another traditional tale about a fur coat:

> A traveller got into a coach on a dark night. He saw only a cloaked figure on the seat next to him, so he said politely,
> "It's cold"
> "Grr" was the only answer.
> "That's a lovely fur coat you're wearing"
> "Grr", was again the only response
> "Would you allow me to stroke it?"
> "GRRR!!!"
> "Help, help, Guard! There's a bear in the coach!" the man cried!
> The bear let down the window and jumped out.

COLLAR

The COLLAR is a piece of fabric, in various styles and shapes, to go round the neck, either on its own or attached to a shirt or blouse. It is a piece of clothing that can be highly emotive or even incite class prejudice! What distinguishes a WHITE COLLAR WORKER from a BLUE COLLAR WORKER?

The white collar is generally understood to belong to **a professional** such as a doctor or a top level business manager – people who don't dirty their hands or their collars. It is admiring if said by the middle classes, but derogatory when the coded reference is to those who don't do 'real' or heavy manual work.

BLUE COLLAR WORKERS are **middle management** or **office-bound administrators** and technicians, but *not* manual workers, who typically wear overalls or other heavy-duty clothing.

RED COLLAR WORKER, a relatively new term, has been coined to describe those **working in the sex industry**. Presumably, this sprang from the association of red with sex and sinfulness rather than from any literal wearing of a collar – unless it happens to be a spiked or bejeweled one worn all by itself.

The word collar seems to be quite ancient, coming originally from the Indo-European root *qwelo* for a wheel. The *kols,* meaning the base of the neck, is something on which the head turns and the Latin *collare* meant both necklace (think of broad metal collars and Celtic torques) and the part of the garment which goes round the neck.

Collar is often used metaphorically. "'Ullo, 'ullo, 'ullo! You're COLLARED!" says the policeman **grabbing a thief by the neck**. Similarly, TO FEEL SOMEONE'S COLLAR means **to lay hold of** and **arrest them**.

To be HOT UNDER THE COLLAR is fairly obvious, as we may feel **a hot, choking sensation** when we are enraged.

Here's a Story. A lady goes to a coat outlet to get a bargain, and tries on some very nice, cheap coats with lovely, deep fur collars. At home that night, she develops a strange and painful sore on her neck and feels ill, so she goes to the local Casualty department. She gets steadily worse, almost at death's door, yet nobody seems to know what is wrong. Then, luckily, a foreign nurse recognises the wound to be the bite of a very rare snake from her own country, and the lady's life is saved by the appropriate anti-venom. Later, retracing her steps on the day she fell ill, the woman arrives back at the rack of fur coats, and finds them rippling as dozens of tiny snakes hatch out in the warmth of the store.

I should point out that this has all the marks of an urban legend, that is, it is almost certainly completely untrue! What kind of snake lays eggs in fur? and how could they be small enough to be missed on making up the garment? But it does make the skin crawl.

COLOURED CLOTHING

The term BLUESTOCKING is still sometimes heard as a term for **an intellectual woman**, not meant as a compliment, and it was used even more scornfully in the past. It comes from the 18th century parties given by Lady Elizabeth Montagu for the enjoyment of genteel conversation and stimulating discussion. Benjamin Stillingfleet was a frequent guest; he wore blue stockings, so groups that started up in emulation of the original soirées wore them too. Since the early 20th century, only women are described as wearing stockings, so that may be why BLUESTOCKING only applies to women. Alternatively, maybe it is only women whose intellectual aspirations are despised.

Here's a colour-related oddity based on a true occurrence. There were once two eccentric gentlemen who set up a charity in their native Derbyshire. Its purpose was to provide cloth so that paupers might make themselves clothes. What generosity! However, there were conditions. Mr Greene and Mr Gray stipulated that the material should only ever be of the colours of their names. I wonder how the recipients felt at being marked out like this, and whether other people avoided shades of green and grey for fear of being thought to be paupers?

A pennant is a long, narrow triangular flag. NATO, the North Atlantic Treaty Organization, flies blue pennants on everything to do with its troops, to distinguish their forces from those of the Warsaw Pact, which are orange. The term BLUE-ON-BLUE is something we would rather not have at all – it refers to soldiers **accidentally killed by their own side**, otherwise euphemistically called 'friendly fire'.

A BLACKSHIRT, literally a man wearing a black shirt, generally has a pejorative meaning for most people born in the 20th century. Initially, such uniforms were worn with pride as being patriotic. Black shirts were donned by the paramilitary wing of the National Fascist Party in Italy during the First World War. Early nationalistic members included landowners and intellectuals, who swore allegiance to Benito Mussolini. The group carried out many acts of murder and intimidation over many years. The extreme right wing views continued to be represented in Britain by followers of Oswald Moseley and the British Fascist Party. A BLACK SHIRT can still be seen as the insignia of **the British fascist party**. I can only add a little British Second world war ditty which sums up my feelings:

> "Whistle while you work
> Hitler bought a shirt,
> Mussolini wore it, Churchill tore it,
> Whistle while you work."

What about the idea of dressing boy and girl babies in different coloured clothing? Most people will quote the cliché, "pink for a girl and blue for a boy". Debate rages hotly on this idea, and there seems little reason for it. Although white is, in some ways, the least practical choice for infants' clothes, because it shows every dribble and stain, white has been popular since at least the 17th Century, probably because the cloth can be bleached, which both cleans and sterilises it. When fast dyes became more common in the late 19th century, the colours were initially reversed, boys being popularly dressed in pink and girls in blue. However, 20th century marketing in the USA and UK ensured the present fashion, though people are beginning to rebel against this dogmatism.

The links between colours for clothing, among other things, and emotions, actions, or ideas is too big a subject for the scope of this book – other than to note that the psychology of colour is fascinating and worth learning about.

Some Monastic orders of Christian Monks take their name from the colour of the simple woollen robes worn to testify to the simplicity of their lives and their identification with the poor and humble. The BLACK MONKS are from the Dominican Order, and the WHITE MONKS are Cistercians. The GREYFRIARS, an order of Franciscans, not surprisingly wore grey habits and were only the second Christian religious house to be founded in Britain, way back in 1225.

It is worth noting that the word SUFI, denoting the followers of a mystical branch of Islam, means literally, **the wearers of woollen cloth**, because they originally refused to wear the rich clothes associated with the Court.

To say THE MEN IN WHITE COATS WILL COME FOR YOU is an insult, implying that the speaker thinks **you must be mentally unstable**, and need to be carted away in a STRAITJACKET [see JACKET] by men from the asylum. Historically, there was very little that could be done for inmates of Hospitals for the Insane, as these institutions were once called. In past centuries, distressed and psychotic patients were usually restrained physically, sometimes with appalling cruelty, and often exhibited to visitors, who amused themselves gawping at the poor souls. It has been a long, slow struggle to destigmatise mental illness and raise the standards in care of its sufferers, with a very long way to go still.

COLOURS ☞ BANNER, BAMBOOZLE and FLAGS

The term colours meaning flags has been used by the English Navy since the early 18th Century. Flags of Army regiments are also known as their 'Colours.' Old-style foot regiments used to have two Colours – the King's

Colour and the Regimental Colours, that acted as the rallying points in battle. It was horribly bad form to allow the enemy to capture your colours, or to allow them to fall to the ground.

TO SHOW YOUR TRUE COLOURS or TO COME OUT IN YOUR TRUE COLOURS are metaphors derived from **identifying yourself** by the flag.

A pirate ship would commonly SAIL UNDER FALSE COLOURS, pretending to be a **harmless merchant ship** or an **ally**, but hoisting the pirate flag as soon as it got near enough to deliver a broadside volley. The 'Jolly Roger', that well known red and black flag with its skull and crossbones, has become known as the symbol of piracy.

In fact, there were quite a few versions of this ensign since the late 17th century. Various suggestions have been made as to the origin of Jolly Roger. 'Old Roger', a name used for the Devil, was represented by a skeleton. Possibly it was a corruption of the name *Ali Raja,* meaning 'King of the Sea', who was a real pirate, or even from the French *'Jolie Rouge'* **– pretty red**, either because some vagabonds and wandering rascals in France dressed in red as a pretence of being scholars, or because so many pirates sailed under a blood-red flag.

If you NAIL YOUR COLOURS TO THE MAST it means that you declare your intention to stick to a course of action or commitments **to the bitter end**, since a flag nailed up couldn't be lowered in surrender.

COUNTERPOINT

The Latin word *culcita,* which meant a stuffed sack or mattress made its way into Middle English as a *quilte*. In Old French *a cuilte counterpoint* was similar, but the outer layers of fabric were stitched through the stuffing at intervals in order to prevent the filling from moving around and bunching up at one end. It was used on top of the sleeper, as is a modern Duvet, and became what we would call a Counterpane, from the French word *point*, a stitch.

{See my book, **Rigmaroles and Ragamuffins** under QUILT}

Pricked holes, similar to a little stitches, in a paper or parchment song sheet, or a series of dots drawn in a music manuscript, were used to show where **a different melody** could be played alongside the main tune, in what is still known as COUNTERPOINT. It is not certain which came first, the music or the quilt, but how lovely if they arrived together, so that a lucky few could drift off to sleep while listening to sweet singing!

COURT CARDS ☞ COAT

What would you understand by COURT CARDS? The very first playing cards were invented in China, perhaps as early as 600 BC – supposedly at the time when the Chinese started to write on individual sheets of paper rather than rolls or scrolls of it. By the 11th century, card games had swept across Asia, and made their way into Europe about 300 years later.

In a pack or deck of cards, the King, Queen and Knave (the last normally called the "Jack" in North America) were called COAT CARDS, because the figures were shown wearing **cloaks**; but by 1690, the word coat had been corrupted, and **Court Card** stuck, as it was just as appropriate.

One might wonder why it took three centuries for cards to catch on in the West. My theory is that it was due to all the compulsive gamblers who refused to admit they were losing. They kept begging for "just one more hand", thereby dragging each game out till the teapot was empty, everyone was exhausted, and a horde of new enemies (or maybe the players' spouses?) were hammering on the gate.

CUFF

The cuff as a band at the end of the sleeve is of unknown origin. In medieval times, these cuffs were often separate strips of linen added as decorative features and sometimes embroidered. It allowed the better-off to alter the look of a dress or coat with a new splash of colour – rather like snapping a new cover onto your mobile phone to keep up with the latest bling.

To cuff someone, meaning to hit them, may come from *kuffen*, a Germanic slang word meaning 'to thrash', in which case, it is not derived from a textile source. However, if I were to CUFF YOU, it might imply **a token blow** with the end of my sleeve, rather than a thrashing.

We say someone is SPEAKING OFF THE CUFF when we mean **without notes**. In the days when cuffs were made of celluloid or even paper, instead of starched linen, an orator could scribble a few quick notes on his cuff to make his speech appear spontaneous. Things could have been tricky for after-dinner speakers who had dipped their cuffs in the gravy.

In the early 20th century, an American who was temporarily short of money might be lucky enough to have a friend with a café or restaurant willing to give him credit. Then he could EAT ON THE CUFF – that is, with just **a quick notation on the end of a sleeve of what was owed**. And if the owner or waiter was a really good pal, that note would conveniently get "lost", so no payment would ever be demanded.

On the other hand [or sleeve], if you didn't make such an arrangement, but simply walked out without paying, the proprietor might shout to a passing policeman, "PUT THE CUFFS ON HIM OFFICER!" meaning **the handcuffs**.

CURTAIN ☞ BED and DUVET

Our word curtain, for a hanging screen of cloth, came about through a confusion between Latin and Greek words in the 4th century AD. Somehow, *cortina* which meant 'a round vessel', got mixed up with the Greek word for 'court', where one might expect to encounter rich fabric hangings. The resulting word, *cortine,* entered English from the Old French, probably thanks to all those posh Norman invaders with their fancy bed hangings. Try draping your window with a clay jug, or even a Ford Cortina car, and you'll soon see the difference!

In earlier times, wives were supposed to 'know their place', and were not meant to say anything critical to their husbands in public. In the 17th century, A CURTAIN LECTURE delivered within the privacy of the great, four-poster bed hangings of the time was the safest way for an angry wife to tell her husband exactly **what she thought of him**. Nowadays, thanks to draught-proofed, centrally-heated bedrooms, we don't need bed curtains, and wives are no longer inhibited from airing their views in daytime and in public. Is that an improvement? I leave your husbands to decide.

To say IT'S CURTAINS FOR HIM is an abrupt way of saying someone is either doomed or **dead**, perhaps deriving from the discreet drawing of bed curtains around a corpse. A CURTAIN-RAISER is **a short, one-act play**, used to start the evening's entertainment in style, while TO BRING DOWN THE CURTAIN can describe **the finale** of almost anything, and comes from lowering the theatre curtain at the end of a play.

It is said that the last words of Rabelais, the Renaissance physician and French comic writer were "LET DOWN THE CURTAIN, THE FARCE IS OVER".

CUSHION ☞ COUNTERPOINT and DUVET

This word is closely linked to QUILT, which was covered extensively in my first book ***Rigmaroles and Ragamuffins*** and is summarised here for completeness.

A Quilt, as any textile person will know, is a soft bag, usually the size of a bed, single or double, filled with feathers, down, silk wool, or a synthetic material – anything soft and warm. It is similar to an eiderdown, but more use as it renders blankets unnecessary, and the bed is much easier to make up. It is related of course to the DUVET. Nor was it only used for bedding; many clothes were made warm by being layered and stitched, and many household items can be quilted.

In the early 17th century the term also developed into COUNTERPANE to describe a covering, for beds, furniture or other objects that were stitched and quilted. The now obsolete PANE meant simply 'cloth'.

We get the word 'cushion' for a bag filled with soft down or rags to ease the hardness of chairs and beds, from the Romans, who notwithstanding their tough reputation, loved comfort. The Latin *culchita* was more of a mattress (see COUNTERPOINT) but also was used for a 'cushion for the hip' though the original Latin word is now lost. A linking word, *coxa* also gave us our medical term for the tailbone, the *coccyx*. It came into English, probably in the 14th century, via a great number of variations, including Old French *coissin, cussin, quisshon* and *cusshin*. Apparently, there were some 70 listed spellings before the word settled down into our comfy cushion in the 17th century.

They all look to be nice, soft words on the page; it's a shame we can't use them all. One of those thin, hard, canvas-worked pads could be a 'quision,' but a really deep, soft, squashy, velvet pillow would surely be a 'cussshhionn'

We might CUSHION A FRIEND, meaning **to help and support them**, as if protecting them from a fall. To CUSHION anyone or anything implies **protecting and comforting them**. You can CUSHION A BLOW, or OFFER A FINANCIAL CUSHION.

What about Curly locks? She is the girl being wooed in the nursery rhyme,

> "Curly locks, Curly locks, wilt thou be mine?
> Thou shalt not wash dishes, nor yet feed the swine;
> But sit on a cushion and sew a fine seam
> And feed upon strawberries, sugar and cream!"

I think I'd go with anyone who said I didn't need to feed the swine, even without the other inducements.

The modern joke item known as a Whoopee cushion, or Poo-poo cushion, which makes a noise like a fart, may have been invented in the 1920's by a rubber company, but antecedents of this rude device made from animal bladders were known from Roman times, and probably invented even before. Human flatulence is possibly the first, and crudest, source of amusement to the race, closely followed by being chased by a woolly mammoth or slipping on a banana skin. Don't tell me woolly mammoths and bananas were never around together. I know! If they had been, those clumsy old mammoths would have become extinct much sooner.

CUTTY SARK ☞ SHIRT

Robert Burns was a very famous Scottish Poet, which of course you already knew, as the Scots make a big thing of eating haggis with neeps and drinking whisky in his honour on Burns Night in late January. But maybe you didn't know that he was the indirect source of the name THE CUTTY SARK for a famous ship, the last and fastest of the Tea Clippers, which now lies in dry dock at Greenwich, England?

In a poem called *Tam O'Shanter*, written in 1791, there is a fearsome witch called Nannie Dee. The poem became very popular, and her likeness was used for the ship's figurehead – no wild-haired hag, but a beautiful, bare-breasted young woman. She was supposed to have been dressed only in a shirt worn since her childhood, hence far too short for her, with erotic effect.

A SARK is a Scots word for **a linen chemise**, to be worn as an undergarment, and CUTTY was the word that showed it to be **cut short**. Tam O'Shanter was moved to cry out "Weel done, Cutty-sark!" which became a catchphrase of his day.

DANDY

One example of this word occurs in a well-known rhyme; here are just a couple of the many recorded verses,

> Yankee Doodle came to town,
> Riding on a pony,
> He stuck a feather in his hat,
> And called it macaroni.
> Yankee Doodle keep it up,
> Yankee Doodle Dandy,
> Mind the music and the step,
> And with the girls be handy.

It was adopted as a song by troops in the American War of Independence, and the earliest reference to words or tune is in 1768 in Boston, the tune being the same as to 'Lucy Locket.'

A DANDY is someone **affectedly trim or neat in his dress**, and derives from a Jack-a-dandy, a fond, 17th century diminutive of the Scottish name, **Andrew**. Fashionable DANDIES of Georgian London were also known as **fops** or **exquisites**, and some were obsessed with dressing elegantly. This evolved to mean **the correct thing**, or something **excellent**, as in the American 20th century phrase THAT'S THE DANDY.

··· Dandy ···

Rather like tootling, to deedle or doodle in the 18th century was to hum or sing without words; one of these may originally have been a dance.

Why macaroni? Well, MACARONI was another word for a fop or dandy in London. Young men who had done the Grand Tour of the Continent felt themselves to be very sophisticated in their tastes and experience. **Macaronic** originally meant a mixture of different languages, particularly a form of verse containing vernacular words, but with Latin constructions. This notion was later applied loosely to any verse form with **two or more languages jumbled together**, such as affected young men might well believe made them sound sophisticated.

While the Yankee Doodle ditty may have originated with the rebels in the American War of Independence, rival British troops supposedly seized on it, invented more disdainful verses and sang these back at the colonists, presumably making fun of the 'country bumpkins' who aped fashionable ways and thought themselves to be DANDIES and MACARONIS.

However, there is evidence that this is not the true origin of the words. They were probably written a good many years earlier, in 1758, when Thomas Fitch, a gentleman from Connecticut, raised a group of local farmers and tradesmen to help Colonel Abercrombie at Fort Ticonderoga, in the French and Indian wars.

This hastily assembled rag-bag of soldiers had guns, but no uniforms. They were about to set off, when Thomas's sister Elizabeth called out, "Wait! Soldiers should wear plumes!" She ran to the hen house and collected bunches of feathers, which the men sheepishly stuck into their battered tricorn hats. When they eventually reached the Fort, a resident Dutch surgeon, Dr Richard Shuckburgh, saw them and was moved to write the original verses.

DISGUISE ☞ DRAG

We are among secrets and hidden things here, very suitably for a concept that came from an ancient source word, now lost, but which gave rise to Old Germanic *wison*. Some of the words descended from it relate to manner, style, or appearance – such as likewise, otherwise, to wit, or in this wise. The next generation of its offspring then shifted a little in Old French to give us GUISE and DISGUISE, in the sense we use today: **to transform, or to change dress and/or appearance**.

Wison also gave us words concerned more with knowing than with seeing, which ended up as 'cousin words', including: wise, wisdom, idea, vision, wit, and wizard.

Travelling actors in medieval times were known as mummers, possibly from Old French *momon,* a mask. This led to *momeur,* 'one who mimes'. Later, these players also recited words and were known as GUISERS or GEEZERS in the UK. They usually wore exotic clothes and **disguised their identity** with masks, or had their faces painted or obscured by their headgear.

After such a performance, rounded out by a boy carrying a broom and a tin mug through the audience for contributions, one might hear this little rhyme from the late 19th century,

> "Here come I, little Devil Doubt,
> With my trousers inside out;
> Money I want, and money I crave,
> If you don't give me money
> I'll sweep you to your grave!"

Wearing clothing inside out in folklore could indicate his 'outsider' demonic status, but is also commonly used as magical protection against spells. The young lad soliciting coins could have been protecting himself since, by tradition, it is dangerous to speak of or imitate the devil.

We are still referring to the old performances of the GUISER when we refer to **an older man** as a GEEZER. It's quite an affectionate term, though rather lower class. If you want to show real appreciation, you could call him A DIAMOND GEEZER. These terms have been around for a long time in Cockney slang – though not their famous rhyming slang – to describe **an admired acquaintance, a buddy, one of the lads**. Linking this intent with the idea of a **rough diamond** – meaning someone who is a bit uncultured and rough but with a heart of gold, probably produced DIAMOND GEEZER. It was used in 2005 by TV writer Caleb Ranson as the name of a series starring David Jason.

DOES MY BUM LOOK BIG IN THIS?

We joke about women's insecurities over their appearance when we use the cliché, "DOES MY BUM LOOK BIG IN THIS?" as it is supposedly uttered plaintively by every woman when she tries on clothes. Not true, of course. Anyway, large buttocks are considered a great sign of beauty in some cultures.

Nobody knows just when the phrase started, but it was a popular catchphrase in the 1990's, when The Fast Show, a television sketch series, featured a recurring character, the Insecure Woman. She took her frenzied questions about how she looked to farcical extremes. The audience may have laughed, but most normal women watching would have to confess that they could relate to her.

Here's a Riddle: What is a woman's ideal dress? Answer; one that makes her look slim and men look round.

DOMINO

This is another word from France, where it signified a hooded cloak worn by priests. It derives from the Latin *dominus*, meaning Lord or master of the house. Wearing a DOMINO conveyed a similar level of authority to priests. Later, it came to mean a hooded cloak with a half-mask worn at masquerades.

For some reason, the name was eventually given to **a game** traditionally played with **small rectangles of ivory or wood** with dots or "pips" denoting numbers. Why the game was given the name DOMINOS or DOMINOES – both spellings are acceptable – is unclear, unless as has been suggested, the winner shouted "Domino!" meaning Lord. Not very convincing – or am I underestimating the excitability of clerics?

DRAG ☞ DRESS and TRAVESTY

In the theatre, of course, it was extremely common in past centuries for men to play female roles. From the rosy-cheeked boy actors of Shakespeare's day to the comic Pantomime Dames of the last hundred years, it is a tradition that has gone on and on. **Men wearing women's clothes** are described as WEARING DRAG. The term goes back only to the mid-19th century, when it was first called **to go on the drag**, or **to flash the drag** – initially only for **transvestites**, but later on, also for **actors playing female roles**.

So, what is the textile connection, given that **to drag** meant originally only to draw or pull something across the ground? Well, it was supposedly derived from the dragging feeling of wearing **a long skirt**, which trails in the mud and also becomes **bedraggled**. It's as good an explanation as any, though the phrase seems unknown before the 19th century, and didn't find its way into print until the 1940's. If you have **a tiresome task** you might say, "IT'S A DRAG", conveying the same sense of towing a ponderous burden.

A man who **dresses as a woman**, either glamorously or outrageously, in a theatrical performance, particularly a solo spot, is sometimes known as a DRAG QUEEN. Such performers are very popular in cabaret and pantomime, wearing heavy make-up, wigs, ostentatious jewellery and sumptuous dresses. What's not to like?

Cross-dressing has, of course, a very ancient history, and there are countless examples in myth, and recorded events. One legendary Greek hero, Achilles, was dressed as a woman by his mother for his protection, because it was prophesied that as a soldier, he would die in battle. His mum gave him a female name and placed him in the court of the King of Skyros as a lady-in-waiting. There he grew among with a group of girls, until lust got the better of him and he made one of his companions pregnant. Odysseus, the great warrior, sussed out the boy's secret, and wanted to sign him up for the Trojan War. He used a never-fail device of throwing into the group a mixture of jewellery and weapons and then sounding a trumpet to make it sound as though the island was under attack. Achilles grabbed a sword and jumped up, ready to fight – which suggests that nature is likely to triumph over nurture in most cases.

In another Greek myth, after Hercules had seriously offended Apollo, the god's Delphic oracle punished the hero by sending him to serve as an attendant to Queen Omphale for a year. While the queen got to swan around in his lion skin, the poor muscle-bound chap had to sit all day in women's clothes, doing women's work such as spinning. The ancient Greeks found this hilarious, and a humiliating punishment for a man. But Hercules had the last laugh, as even in his dear little girly chiton, the Queen of Lydia fell in love with him and freed him so they could marry.

Historically, to dress a man in women's clothes was seen as demeaning. Once, it is written, a mischievous girl in a Palestinian Arab story triumphed over a boy and his forty brothers by a series of clever tricks. In one, she gets him drunk and then shaves him and dresses him in her clothes. In the end, she relents, turns aside his anger with one last trick, and then marries him, also providing suitable brides from among her friends for all his forty brothers. It's a neat reversal of the more usual plot of the fairy tales we have adopted in the West and Disneyfied, where the men are the more active protagonists or heroes, and the women more commonly seen as passive.

In 19th century Wales, the Rebeccaites used cross-dressing with a serious purpose. Actually, they were tenant farmers who banded together against the oppressive turnpike tolls, parish rates, church tithes, and property restrictions of their day, which were squeezing them into starvation. They dressed up in women's clothes with faces blackened so that they would not be recognised, and assumed a collective name taken from the biblical Rebecca, daughter of Laban, who blesses her with the words, "Let thy seed possess the gate of those which hate them." The Rebecca riots of 1839 and 1842–43 in parts of South Wales were crushed in violent fashion by the military.

Cross-dressing is also a frequent motif in folktales – particularly girls dressing up as men to get along in a man's world. The song Sweet Polly Oliver has a young woman going as a soldier to be near her sweetheart, and nursing him when he was wounded. In every war, there have been real cases of women enlisting, often only discovered after the death of the supposed man.

Not quite so sweet is a story from Surinam, in which a wife plays a trick on her husband, who is about to catch her with her lover. She quickly dresses the lover as a woman and pretends to read out a letter in which her sisters tell her they have turned into men. Supposedly, her lover is one of them. As the husband can't read, he believes her deception, and she gets away with it. Moral of that? Learn to read, and be careful who you marry, I suppose.

DRESS ☞ DRAG and TRAVESTY

TO DRESS has been linked with CLOTHING for so long that it would be impossible not to look at the many uses of the word and sayings derived from this verb.

The origin of our word DRESS was the Latin *directus,* which meant 'to put straight', and also gives us 'direct'. In Old French, the word became *dresser,* or *dresseur,* as someone who 'dressed' food – that is, prepared it, decorated and displayed it, and also put it out on plates. By the late 14th century, the word *dresseur* or *dressoir* had migrated from the **servant who prepared the food** over to the **table** where the work was done. It came to mean a different piece of furniture where plates and utensils could be kept conveniently close in the kitchen or dining room. It usually consisted of legs or bun feet supporting a cupboard with doors, a surface at table height, and above, open shelves up to the top of the unit, where crockery could be displayed and kept. A DRESSER is still used for a **sideboard for food preparation**.

The word evolved various meanings related to **making preparations**, getting things **lined up** or **putting them right**, so we have DRESSING troops by **preparing them in columns**, or DRESSING a roast or turkey for the oven, as well as TO ADDRESS someone, or a STREET ADDRESS where we live. If we want to talk about horses, we might discuss DRESSAGE, which is **training our equine friends to perform balletic manoeuvres**.

Although semantically quite different, it is interesting that the Welsh employ the word *tidy* in a much broader sense than the English – more akin to the original use of dress – to mean not just **neat in appearance**, but **right and proper, straightened up** and **well organised**.

If you are DRESSED TO KILL you are out partying, and **hoping to make a big impression**. When you are on the razzle-dazzle, someone might say you are DRESSED UP TO THE NINES. No one is sure how this latter odd phrase came about, but there is a suggestion that, because the plural of 'eye' used to be 'eyene', it conveyed a superlative meaning of being sartorially adorned **right up to the eyes**.

A DRESSING DOWN is **a severe reprimand or telling-off**, although it seems to have been in use since the 15th century, without the addition of down, and meaning a beating, a drubbing, a physical chastisement – generally with the sense of being well deserved.

However, TO DRESS DOWN means **donning unostentatious clothing** for an occasion. A new phrase, DRESS-DOWN DAY has been coined for **a day at work when casual clothing can be worn;** often it's a Friday preceding a relaxed weekend or general holiday, but it can also be used colloquially to mean any sort of **informality**.

So, to be fair, after dressing down, we must mention DRESSING UP. This is what children do, putting on their parent's clothes and accessories while **pretending to be grown-ups**. This is lots of fun, but for goodness' sake, any children reading this, please don't raid the wardrobe without permission! If you do, your parents will find their best clothes and makeup spoiled when they, too, want to DRESS UP. On second thoughts, you might actually be doing them a favour, since some of those social events can mean squeezing into one's best **clothes** and spending a dreadful evening with boring companions. Not much fun!

You might say jokingly, I'M ALL DRESSED UP AND NOWHERE TO GO. This was the title of an American song, written by Benjamin Hapgood Burt in 1913 – "When you're all dressed up and there's no place to go". In fact, another songwriter, George Whiting, had written a slightly different set of words with the same title a year earlier. The later ditty was popularised by a stage comedian, and a few years after, a journalist used the phrase to describe Theodore Roosevelt as "All dressed up, with nowhere to go", when the candidate suddenly withdrew from his Presidential nomination. It caught on as a popular saying and has now become a cliché.

When measuring a gentleman for trousers, a tailor asks, "Which side do you DRESS, sir – left or right?" meaning **to which side of the crotch does their manly tackle incline**, because the garment's construction must allow for a bit of extra space on that side.

WINDOW-DRESSING can be meant literally to describe **displays in a shop for passers-by**, but is also figurative, meaning **to flaunt your skills** or attributes and be very up-front about yourself – usually with the mocking sense that **there is nothing behind the superficial show**.

An advertising headline that also became a cliché is now used sarcastically. 'What the well-dressed man about town is wearing' has given birth to a series of ironic slogans describing scruffy tramps or builders' labourers. So today, we might as well say, WHAT THE WELL-DRESSED BAG LADY IS WEARING.

The phrase MUTTON DRESSED AS LAMB – meaning **dressing in a fashion younger than is appropriate for your age** is really clever word-play on the other sense of dressing, i.e. preparing a joint of meat. You might try to **disguise scrag end** of a tough old ewe as tender young lamb (or to convince the world you are still in your prime), but sooner or later, someone will see through your ruse.

DRESSED UP LIKE A DOG'S DINNER is a disparaging way of describing someone **overdressed and flashy**, whereas DRESSY suggests the opposite: that **you are fittingly dressed in a formal style**, and generally interested in dressing well.

Another non-compliment for **someone wearing flashy and inappropriate ornaments** is ALL DRESSED UP LIKE A CHRISTMAS TREE. How easily women seem to slip into snide 'compliments' which are really each a little cat-scratch.

Personally, I love showy bling, and I can't see it as a problem. And as I always say, "Why wear one piece of jewellery where six will do?"

DUDS

Why would you refer to your clothing as your DUDS, as if you were carrying round a DUD First World War shell? I could suggest, "because it's better than carrying round a live shell", but that would be just silly.

The word DUDS, in the sense of clothing, goes back a long and uncertain way – perhaps to the German *dudel*, meaning coarse sackcloth. As it is usually a slightly disparaging term, denoting clothes that are **old and shabby**, it may come from a term, known since 1500, referring to a beggar in ragged clothing. In the 19th century, the word began to carry the idea of someone or something **useless**, leading to the use of DUD in the First World War to describe shells which **failed to explode**. Now we utter it colloquially for anything which **doesn't turn out well**. I think one might very reasonably feel that a shell which didn't explode has turned out extremely well – especially if you are standing anywhere nearby.

DUVET DAY ☞ BED, CUSHION, and QUILT

A DUVET is a soft bag, a bit bigger than the size of a bed, filled with feathers, down, silk wool, or a synthetic material, as with its close relation the CUSHION. Indeed, the word derives from 'down' meaning the soft under feathers of a bird, from an old Norse word *dunn*. The DUVET came to the UK from Europe and was originally known as a Continental quilt, and after dispelling the Americans' traditional suspicion of anything not invented within their borders, they took to DUVETS with enthusiasm. While similar to an eiderdown, the DUVET is more practical, as it renders blankets unnecessary. A bed is much easier to make, too; a quick shake and the duvet looks smooth and inviting again. In fact, I think I'll crawl right back in again and have a snooze.

A company in the UK invented the idea of the DUVET DAY which is **an agreed number of days**, often four, **that can be taken off in any working year without notice by the worker** – sometimes within, and sometimes extra to, the contractual holiday allowance. It means that, if someone is hungover, tired, dealing with a small domestic crisis such as a broken water pipe, or simply needs a break, they don't have to pretend to be sick – 'Throwing a sickie,' or 'Going on the sick,' as we say in Wales.

It was said to have been started in 1997 and to have markedly improved both company morale and attendance rates. In recent years, DUVET DAY is sometimes used in a light-hearted fashion by others, who are not workers, to describe a day when they don't get dressed or go out, but **snuggle up on the sofa under a duvet** to watch rubbishy TV, eat biscuits, and ignore their responsibilities – definitely my sort of day!

EAT MY SHORTS

Where did Bart Simpson of the animation series, 'The Simpsons', get his phrase EAT MY SHORTS?

According to Nancy Cartwright, who does the voiceover for the iconic delinquent, she used it in a spontaneous moment of inspiration in 1989 and it stuck. It was dropped for a few years, presumably on the grounds of taste or due to complaints, but came back in 2000.

Shorts in this American context means underpants, not shortened outer trousers for younger children or sportswear. There is some controversy as to whether this was already in use in 20th century US high schools, but Bart Simpson certainly popularised it worldwide.

There is a humorous equivalent to EAT MY SHORTS in KISS MY CHUDDIES – chuddies is Hindi for underpants, used as a catchphrase from the BBC TV comedy show of 1998 onwards. It was called, 'Goodness Gracious Me' and was based on the experiences of immigrants to the UK from the Indian subcontinent.

Bart's catchphrase is linked with **mooning, in which the buttocks are bared**, as can be seen in the cartoons about him. Any gesture that involves **showing the underwear or the naked genitalia** tends to be taken a sign of defiance and disrespect. When the explorer Verrazano voyaged along what is now southern Maine's shore in 1524, he was insulted by disdainful natives. In his letter to the King of France, he wrote, "We found no courtesy in them, and when we had nothing more to exchange and left them, the men made all the signs of scorn and shame that any brute creature would make." This disrespect seems to have included baring their buttocks and laughing at him. Poor chap. But at least he eventually had a nice American bridge named after him!

For many millennia, anything to do with bodily functions, especially reproduction, has been regarded as simultaneously sacred and shameful. The naked genitals signify life itself, and in some of the Eastern European countries an annual ritual took place in which peasant women would expose themselves to the growing flax – a form of sympathetic magic designed to encourage fertility of the flax harvest.

ESCAPE ☞ SCAPEGOAT

A less immediately obvious clothing word is ESCAPE, from the Latin *ex* meaning out of, and *cappa*, which was another Roman word for a cloak. It provides a picture worthy of a theatrical farce where a man is nearly caught by his cloak, but slips out of it and runs away naked.

TO ESCAPE now means a variety of things suggesting **getting away**, including to get off safely, avoid threatened evil or gain one's liberty by flight – even if it's a holiday escape to an idyllic resort.

An ESCAPEMENT is **the regulator in a watch**, and it keeps time ticking accurately by allowing the timekeeping gear train to advance or 'escape' by a precise amount. The first one to be invented in the 1650's revolutionised watchmaking.

An ESCAPADE is **a prank**, a mildly naughty activity by which you escape from strict propriety. It might be carried out by a SCAPEGRACE, **someone whose reckless or wayward actions place him outside grace; a rascal or scamp.**

··· *Escape* ···

THE SCAPEGOAT is really an ESCAPEGOAT, and was a real animal employed in the ancient Jewish tradition; **the sins of the people were transferred** to this luckless beast and it was then driven out into the wilderness and allowed to escape, taking the sins with it. We use the term to describe any **innocent person who is blamed for the wrongdoing of others**.

FILLET ☞ LEATHER

This is very definitely a 'thread' word, coming as it originally does from the Latin *filum* for a thread of flax. It then made its way to the UK via the Normans and Old French as a generic FILET, usually meaning a strip of leather, cloth, twisted threads or even a thin, decorative band of precious metal worn on the head to secure one's hair, as the Roman ladies did. In English-speaking countries, its Franglais source probably accounts for variable spelling of the word as either A FILET or FILLET.

There is much discussion too of the the ways of pronouncing it – as the French *fee-LAY* or the more down-to-earth anglicised *FILL-ett*. There isn't even any online agreement as to which is the more appropriate to use. Where there is no consensus, I think only a food snob or a Francophile trying to show off would really care!

So, A FILET or FILLET in Haute Cuisine [French for Posh Nosh] is **a small strip or piece of boneless meat or fish**. If you are using a filet as a band around your head, I suggest something more acceptable dress-wise than a sliver of fish or meat. However, Madonna famously dressed up in beef steak for one concert – sort of a "filly con carne".

Of course, a filet can also be used as a strap to bind things together or fasten your sandals; it's a ribbon, a tie, a thong, or a lace. In fact, A FILLET is even an architectural term meaning **a strip to round off a structural corner**, or as **part of other mouldings** – you know, that thingy up there on the wall between the thingamabobs. It's a useful all-purpose word really, for everything from your main course at dinner to a builder's finishing touches. Bravo for filets, however you pronounce them!

FLAG ☞ BANNER and COLOURS

The origin of the word flag is obscure, but in the 16th century, 'flag' as a piece of cloth on a pole, or on its own and used as an emblem, were both well known meanings. So was the verb 'to flag' suggesting 'to droop' or 'to get tired'. All these senses have survived into our own times. By the way, the yellow Iris flower known as a 'flag', and a 'flagstone' as paving are probably not connected – unless, like me, you forget where you've planted things and lay your garden path on top of the poor little bulbs.

Britain's navy had a tradition of sending a fleet of ships with their flags flying **to represent Great Britain** in other ports around the world, either taking diplomatic envoys in times of peace, or a convoy of armed vessels to intimidate 'Johnny Foreigner'. President Roosevelt dispatched the American Great White Fleet around the world with a similar purpose – and indeed, most maritime powers do the same.

The terms TO SHOW THE FLAG and TO FLY THE FLAG, have evolved a much more general sense of **taking pride in one's own country**, company, or any band of colleagues, and a willingness to stand up for them. You might also be prepared to RALLY ROUND THE FLAG, a term derived from warfare, where you would literally **make a last stand** around your chieftain's banner. These days, it's applied figuratively to an exhibition of loyalty to any group or cause, even if it is sometimes used ironically.

Although flags have always been important in the Army and in Civic life, their greatest value used to be for transmitting important messages at sea. Before wireless telegraphy – and later, electronic communications – visual signals were often the only way for messages to be passed between vessels or from ship to shore during a battle. The meaning of all the different Naval flags is not relevant here, but it is a matter of pride that textiles, once more, played a key role.

Probably the most famous signal in our history was the one with which Admiral Lord Nelson inspired his men as they sailed into battle at Trafalgar, "England expects that every man this day will do his duty". The story is well known, but perhaps not so many people know that Nelson originally intended to say "England Confides That Every Man this day will do his duty." Hardy, the ship's captain, whose job it was to ensure everyone would see and understand the signal as quickly as possible, explained to his boss that it would take fewer flags to say "expects", and Nelson agreed. See what a good editor can accomplish?

The FLAGSHIP is **the ship of the fleet on which the most senior Admiral or Vice Admiral is sailing**, and it displays his pennant, thus we use this term figuratively when we describe a group or organisation showing **leadership excellence** as A FLAGSHIP for the rest.

Anyone achieving success in an endeavour, such as top marks in an exam, can be described as reaching their goal WITH FLYING COLOURS. In similar vein, ALL FLAGS FLYING can suggest **defiant pride**, when a ship is sunk in warfare, or in happier times, after success in battle, when the fleet sailed home with all their flags fluttering to show that they had been **victorious**.

By contrast, no one likes having to admit defeat. In a sea battle, there was more than one way of surrendering, and some have become metaphors to show that you are **backing down, giving in, and eating humble pie**. For example, TO LOWER THE FLAG or TO STRIKE THE FLAG and THE FLAG'S DOWN – all mean one thing: **we submit**. However, in other contexts, such as **the start of a race**, the phrase THE FLAG IS DOWN is a good thing, because it means something is finally happening, and we are allowed to order hot popcorn and sweets to munch.

When illiteracy was common, flags were sometimes used to advertise entertainments or wares on offer. During Lent, a traditional time of abstinence from pleasures, there was an old custom of closing theatres in London and removing the flag that normally hung outside. So saying THE FLAG IS DOWN can mean **'nothing doing'**, sometimes in a vulgar context.

In the 19th century, the opposite term, THE FLAG IS UP provided a euphemism for **menstruating**, as a FLAG was slang for **a sanitary pad**.

A FLAG OF CONVENIENCE, while apparently legal, sounds like a bit of a sharp practice. Not infrequently, **a ship is registered in a foreign country** in order to get more favourable regulations or avoid certain operating costs.

A flag flown upside down is a FLAG OF DISTRESS proclaiming that the vessel is **in imminent and serious danger**. Other flag messages have become notorious, too, such as the YELLOW FLAG for **disease aboard requiring quarantine**. And when a ship RUNS UP THE WHITE FLAG, it signals its surrender or neutrality; either way, it's not fair firing your cannons at this one.

A bit of modern management-speak about **brainstorming ideas** is "LET'S RUN THIS UP THE FLAGPOLE AND SEE WHO SALUTES." This phrase was known by the 1980's, and started (unsurprisingly) in the USA. You might also say "I'D JUST LIKE TO FLAG UP THIS IDEA," when **you are anxious to know how a suggestion will be accepte**d, as if you were raising a little flag (and hoping that it wouldn't be shot full of holes).

We still speak of FLAGGING DOWN a bus to get it to stop. In the days when trains chuffed along very slowly, a railwayman on a platform or standing beside the rails would wave a flag to stop a train. The Locomotive Acts, later called the RED FLAG LAWS, required a pedestrian carrying a lantern and waving a red flag to walk ahead of any early motorcar to warn bystanders, carriages and those driving animals likely to be spooked of its approach.

A red flag is an internationally-recognised warning, and we talk about FLAGGING UP DANGER metaphorically. THE RED FLAG is also the anthem (still sung), of the International Socialist Party, so TO FLY THE RED FLAG or KEEP THE RED FLAG FLYING, indicates **your allegiance to this group**.

The **Blue Peter** is a blue flag with a white square in the centre, hoisted when a ship is ready to sail; it also stands for the letter 'P' in the International Code of Signals, hence 'P' for 'Peter'. Originally, the flag was used to ask for **a signal to be repeated**, and **Peter** is a corruption of **repeater**. BLUE PETER became the title of a highly successful children's show aired on the BBC since 1958.

In the UK, you can buy a symbolic little paper flag on a pin on relevant FLAG DAYS to support a particular charity, but America's official FLAG DAY is June 14th, the anniversary of the day in 1777 when the Stars and Stripes was adopted as the American flag.

FLOUNCE ☞ FRIPPERY

If we are teased, we might FLOUNCE out of the room, and express a kind of **pretend annoyance**, as if we were swishing the frilled edge of a dress or petticoat. Perhaps we should also mention a FURBELOW, which started out as a flounce and then became the word for **any showy ornament**. And let's not forget FRILLY. This started out to describe a frilled edging, but now means **a showy and useless embellishment** of any kind.

FLYING BY THE SEAT OF YOUR PANTS

In the early days of aviation, when instruments were not very accurate and a pilot had to fly mostly **by instinct**, he would be described as FLYING BY THE SEAT OF HIS TROUSERS – a phrase first recorded in Britain, which flew across the Atlantic without a hitch and became FLYING BY THE SEAT OF HIS PANTS. We still use the latter phrase to suggest any sort of **instinctive or intuitive actions**, even though in the UK 'pants' means underwear. It could get terribly chilly flying in one of those rickety biplanes dressed only in one's knickers. I vote for longjohns!

FRIEZE

The decorative FRIEZE that you might stencil around your walls was originally an embroidered cloth from Phrygia in Asia Minor, then famous for its embroidery. Classical Latin described it as *Phrygium*, meaning 'work from Phrygia', and this evolved into *frigium* and then *frisium* in medieval Latin.

These terms then still denoted 'embroidered cloth', but by the time English inherited the word from Old French as in the mid 16th century, *frise* had come to mean a **decorative band along the top of a wall**. We also inherited **fringe** from the same source.

A little confusingly, the natives of *Frieseland*, now a part of Holland, gave the word *frieze* to a tough type of **woollen cloth** for which the area was famous in the 17th Century.

FRIPPERY ☞ FLOUNCE

This nice word trips easily off the tongue but is not often heard now, which is a pity. It is very appropriate to describe **unnecessary ornamentation** on an outfit as FRIPPERY, since it came from a textile origin, but now it can signify **empty display** in **dress**, **architecture** or even **language**.

The word came originally from the French, as a 16th century word, *frepe*, which became *freperie* for a rag. It could also refer to **secondhand clothes**. When FRIPPERY is used in its original sense to describe someone's clothing it joins the company of such pleasurable words as trifles, gewgaws, tawdry, and frivolous.

FROCK ☞ DRESS and ROBE

My own view is that using FROCK for women's wear is a little bit formal and old fashioned, but I expect a howl of protest from people devoted to the term. The FROCK started out as *froc* in Middle English and denoted a long habit – rather like a tunic but with long, loose sleeves – as worn by monks and the clergy. However, the word originated much earlier in Old High German, where the *hroc* was a mantle or coat; it then migrated to us via the Old French *froc*, which was only being adopted for women's wear in the 16th century. English also keeps this link alive via the frock coat, worn by men at formal affairs such as visiting Buckingham Palace to be knighted, or to weddings if you happen to be the groom, best man or bride's father.

If we were to talk about UNFROCKING A CLERGYMAN [Go on – I dare you!], we would be harking back to the original froc of his robe; hence, TO UNFROCK him means the token **removal of his right to wear 'official' garments** after he is dismissed from his priestly office for some dreadful behaviour.

From at least the 16th to the early 20th century, the FROCK was a popular women's garment, since its colour, fabric and ornamentation could be stylish, while the garment itself was generally loosely fitted and casual – more comfortable than most 'dressy' clothes. The term was used in Australia, but there, TO FROCK UP was **to dress formally** for a special occasion.

The FROCK, SMOCK or SMOCK-FROCK was a long, loose overshirt in tough, closely-woven material worn by farmworkers, waggoners, shepherds, and sailors. Some had buttons and others were pulled over the head; the full sleeves, breast and back were gathered into tucks held in place with cross-stitch or honeycomb embroidery – a style still called "smocking". Later, this form of garment was adapted as clothing for babies, girls and women.

Men of higher rank and class might wear a frock coat, which was double breasted and had long, somewhat full skirts. These continued right into the 20th century, the skirted part later being cut away or left only at the back as tails, giving birth to the gentleman's formal tail coat for evening wear. It wasn't very safe to be worn on a bicycle because of the chain, so when these handy vehicles were motorised, a special FROCK MOTOR was invented that had **no moving parts sticking out** to catch the unwary rider's clothing.

FUR COAT, NO KNICKERS

This lively expression describes **a woman who is no better than she should be**, as do the similar phrases RED HAT or RED SHOES AND NO KNICKERS. The phrase FUR COAT, NO KNICKERS suggests that by wearing no knickers, in other words **by being available sexually**, one has acquired the trappings of success – or perhaps just confirmed tartiness: a fur coat and flashy red garments.

Why red? Well, in the West, this is the colour traditionally most associated with immorality, as in Scarlet Woman and Nathaniel Hawthorne's tale about an adulteress called 'the Scarlet Letter'. In the mysterious East, however, particularly in China and India, red symbolises good fortune, long life, power and happiness, and is often chosen for the bride's wedding garb.

Our English sayings are usefully ambiguous, in that they may also refer to someone in a less expressly sexual way, simply hinting that the person is **all expensive show** on the outside, but **lacking taste or social grace**s. "She's got everything except breeding!" the older generation would say.

GALA

Sometimes you might be all dressed up and actually have somewhere special to go, such as a GALA. Hooray! The word originally meant **an especially fine garment presented as a gift**.

This came from the Arabic *khil'a*, and it was adopted as GALA by the Spaniards, Italians and French, who all have the reputation of knowing how to enjoy themselves. It even found its way into English. It has lost both the sense of presentation and of clothing since the late 18th century, and all that has been left is the idea of a GALA as **a festive occasion**.

GARTER ☞ SOCKS

I'm afraid the only metaphor I know using this word is a gory one. I'LL HAVE YOUR GUTS FOR GARTERS expresses a horridly violent threat **to disembowel your enemy and use his guts to hold your socks up!** Not nice at all and, in any case, not nearly as effective as elastic.

The garter is now practically obsolete in these days of self-supporting socks. An unrecorded source word from Gaul meant leg, and was related to the Welsh for leg, which is *gar*, so that is the probable source of garter. In Old French, it became *garet* or *jaret* meaning "the bend of the knee", and a *gartier* was a band just above or below the knee tied firmly to hold up your hose.

We may not have very much use for them now, but garters in previous centuries were a supremely important erotic focus, as can be seen by the countless poems, plays, pictures, letters and diaries in which they occur, and by the lustful thrills caused by the action of a woman tying or untying her garter. As the Cole Porter song has it, "In olden days, a glimpse of stocking/ Was looked on as something shocking/ Now, heaven knows, anything goes."

Garters played a very important part in folklore, particularly at weddings with their associated fertility rituals; they were often untied ceremonially not only by the groom, but also by attendant young men who would fight to wear the still-warm ribbons as favours. These trifles symbolised the groom carrying off his bride and undressing her, so you can imagine the bawdiness of many a country wedding. In the 17th century, white ribbons might be attached to garters, making them easier to remove. These were also called garters, and could be detached and given away after a wedding for luck.

If you had CAST YOUR GARTER, it meant you had **secured a husband**, and some brides still wear a token, frilly garter under their wedding dress.

Bridal happiness is one thing, but not everyone had such a fate. Think of wisewoman, Ann Greene, who was convicted of witchcraft at a trial in 1654. She had cured John Tatterson's earache by removing her garter and crossing his ear with it. Was he grateful? Not a bit – he became her chief accuser!

I find it astonishing that there are so few figures of speech that relate to garters. It is as if the little accessory is so potent that the human mind, particularly the male human mind, simply can't bear to move away from the reality even far enough to create a metaphor.

However, we can certainly illuminate the cultural significance of garters. The legend about the founding of the British Order of the Garter, is well known – at least in Britain. It is said that one of the ladies at court accidentally lost her garter while dancing. When the courtiers tittered at the poor lady's humiliation, King Edward III picked it up and bound it about his own knee uttering the chivalrous words, *"Honi soit qui mal y pense,"* meaning shame to he who thinks evil of it.

The lady concerned has been variously named as the Countess of Salisbury, Joan the Fair Maid of Kent, or Queen Philippa herself, and it has been asserted that it was not a garter but the belt and rags with which menstruating women tried to deal with their monthly flow in earlier times, which would give more point to the shame, since even today menstruation is still one of the great unmentionables. However, it's difficult to believe that the monarch would have bound such an 'unclean' object around his leg, but that may be the point of the story?

Whatever the facts of the matter, King Edward III went on to found a special order of chivalric knighthood limited to Royalty and a very few commoners, and soon after, Ladies of the Garter were honoured with the Order. It is no longer a simple garter and one would not use such an amazing honour to hold your socks up. It has been developed over the centuries and now consists of an ornate collar with linked tudor roses, a badge or pendant with St George on horseback fighting the Dragon, and additional stars and ribands to wear on less formal occasions. The ORDER OF THE GARTER is still awarded annually "at the sovereign's pleasure" that is, as the personal gift of Queen Elizabeth II, and is pinned onto the gorgeous, velvet mantles of the knights by her each St George's Day. It is the highest order of knighthood anyone can receive, and is worn with pride by the Knights Companion and Ladies Companion, as recipients are known.

In the mid-18th century a comic verse was written in which was the exclamation of surprise, "OH MY STARS AND GARTERS!" – referring to the Order, not the stocking support – first appeared in print. The poem was called 'A Journey to Oxford':

> "Supper at such an hour!
> My stars and garters! who would be,
> To have such guests, a landlady."

A landlady is not merely someone who owns and rents out a house, of course. In this context, she is the publican, or owner of a pub, serving her guests at an inconvenient hour of the night. Stars, garters and landladies were jocularly linked in common parlance thereafter. Indeed, in the mid-19th century, a pub in Birmingham was named the STAR AND GARTER, and it later became a popular music hall and entertainment venue.

A SLEEVE GARTER is worn by white collar workers **to keep their sleeves clean** by holding them in a small elasticated band. These originated in the 19th century.

Cross-gartering is for ever linked with poor, mocked Malvolio in Shakespeare's *Twelfth Night*. He was tricked into wearing exactly the sort of thing his employer hated, and his tormentor, Maria the maid, described in glee the effect – "He will come to her in yellow stockings, and 'tis a colour she abhors, and cross-gartered, a fashion she detests!" Poor Malvolio. He wore on each leg a wide silk ribbon twisted round itself once at the back and tied in a flamboyant bow, not something criss-crossed all the way down like a maypole, as it is sometimes pictured. [Twelfth Night, Act 2, scene 5. 1.192–4, as described in Chambers Dictionary]

In a poem to a lady, an 18th century beau giving fashion hints tells her:

> "Make your petticoats short, that a hoop eight yards wide
> May decently show how your garters are tied,"

Hmm. I'm not sure 'decently' is really what he has in mind.

Garters do feature in a non-erotic way occasionally. After all, they were a practical piece of clothing for much of their time. In an Irish fairy tale, a young man catches a leprechaun, and forces the little creature to take him to his treasure, which is under a big "*boliaun*" or thistle. The boy puts his red garter round it while he goes back for a spade, letting the leprechaun go, after making him swear he won't remove the mark. When the lad comes back every boliaun in the forty acre field has an identical red garter on it.

GIRDLE ☞ LOOSE WOMEN and ZONE

"I'LL PUT A GIRDLE ROUND ABOUT THE EARTH IN FORTY MINUTES" says Puck, in Shakespeare's A Midsummer Night's Dream, to show how quickly he can go on his errand and return.

The word GIRDLE comes from *gurd*, a prehistoric word meaning surrounding or encirclng; it gave us the noun *gyrdel* in Old English, later girdle, for a belt, sash, or other tie to go round the waist. It also gave us the verb TO GIRD for putting on a belt or binding, but it is also used in a metaphorical sense meaning to ready yourself for action in any context. We might still talk of

GIRDING OURSELVES UP **to tackle a problem** or GIRDING OUR LOINS **ready for action**, though it's a little old fashioned.

King Edward III was so inspired by the legend of a knight called George that he made him England's patron saint, even though he never set foot in England – if indeed he existed at all. His story was brought back from Palestine by the Crusaders. It is typical of a story type in which the removal of the woman's GIRDLE by the man – with definite erotic overtones – often signified the taming of the wild and uncivilised male by the more refined female.

A dragon in Asia Minor laid waste to the land and demanded a human sacrifice every day for food. Eventually, it was the turn of the King's daughter to be handed over, but George, a Christian Knight, hearing of this monster, vowed to kill it. After fighting long and hard, he made the sign of the cross to seek heavenly assistance and was able to run it through with his sword. Even so, he could not slay it. In ancient legend, a virgin often has the power to subdue wild creatures, and sure enough, the freed maiden tied her gold silk girdle round the dragon's neck and led it to the market place. There George severed its head, thereby releasing the kingdom from its curse. Seems a bit hard on the poor beast, which had already been subdued. Couldn't she have offered it Anger Management, Psychotherapy and Sedative medication and kept it as a pet?

There is often some confusion between the words GIRDLE and griddle, as to the Scots A GIRDLE is something you cook cakes on, but the word 'griddle' itself originated in terms meaning a grill for cooking on, more related to our 'grate' and not to textiles.

GLASS SLIPPER

In real life, a Glass Slipper sounds extremely uncomfortable, not to say dangerous. One stamp of an impatient little foot and your Handsome Prince will be picking out the splinters.

The well known story of Cinderella, a lowly kitchen maid who charms her Prince at the palace ball by appearing all gussied up in a beautiful costume and GLASS SLIPPERS, has delighted hearers for centuries – even to the extent that the 'Cinderella story' is almost synonymous with Fairy Tales. Or is it? Folklorists have collected literally hundreds of versions of Cinderella from all eras and all corners of the world, containing all the traditional motifs. There's everything from classical Grecian sandals to furry Inuit mukluks, and Chinese tales of concealed and later discovered heroines in which articles of clothing, especially footwear, are key components. THE GLASS SLIPPER is only one of these items, deriving from the Royal Court of 18th century France.

Why then is the story so popular? Well, there are a great many slang expressions and euphemisms for sex or intimate body parts that also relate to shoes – particularly the action of sliding the foot into a shoe or boot. What is the popularity of Cinderella's glass slipper if not a veiled allusion to the Prince's wish, shared by most men, of ensuring that he finds the right sexual partner, the girl who literally best fits his…(ahem) requirements?

Of course, the slipper in the 18th century version was not what you think. GLASS, which is *verre* in French, was a mistranslation of the homonym *vair*. This was **a type of squirrel fur** once used to trim clothing – including elegant shoes or slippers, and it clearly accentuated the story's sexual symbolism.

GLOVES

Not surprisingly, gloves are linked to words for hands. The probable lost root word was *galofo,* in which *ga* is a prefix and *lofo* means hand in the prehistoric Germanic tongue.

To say you are HAND IN GLOVE with someone means you are very **close and cooperating**. It could also suggest that you are **partners in something nefarious** or illegal. The phrase started out as HAND AND GLOVE and took its present form in the 17th century.

TO TAKE OFF THE GLOVES or do anything WITH THE GLOVES OFF means the same as 'no holds barred', and is a fighting metaphor. To STRIKE SOMEONE WITH YOUR GLOVE or to THROW DOWN THE GLOVE or the GAUNTLET – an older word for a chainmail GLOVE – was the way you could **challenge a rival to a fight** or a contest of strength and skill in former times. The word *gantelet* was French, and in medieval times, to throw this item down in front of another knight was a **definite insult** and an invitation to him to TAKE UP THE GAUNTLET and **agree to fight**. Such potentially fatal duels – usually with pistols or swords – were only banned by law in the 18th century.

Because of their symbolic representation of the hand, GLOVES have been extensively used to convey significant messages visually. Sometimes these symbols could be confusing; a bare hand, especially a fist, was usually a sign of enmity; however, in the presence of the King, it was essential to remove your GLOVES – probably to show you weren't concealing a dagger. GLOVES were also given as signs of peace between countries, signifying the hand of friendship, and as love tokens from ladies to their beaux.

A glove could also stand for a pledge, conveying a person's hand on a bargain. Also, given their association with competition or fighting, THE GLOVE'S UP meant a sort of **truce within a battle**. This phrase comes from medieval times, when certain towns were permitted by Royal Charter to hold fairs that were both big commercial and social occasions. Special laws applied to the period when a fair could be held, and because such occasions always attracted a motley crowd of pickpockets, gamblers, mountebanks and beggars as well as honest folk, there were rules about how good order should be maintained – hopefully, without restricting the cash-generating potential of such a lively gathering.

At a time when few people could read, the glove became a symbol of Royal approval, so when a large glove was hoisted high atop a long pole for all to see, it signalled the start of the fair. Hurrah! One such, the Chester Glove, is a carved wooden replica of a hand, and it is now in the possession of the Corporation of Liverpool. Although dated 1159, it is probably a 17th century replica of the original. Afterwards, when the glove was taken down, the idea was that people should disperse in orderly fashion. Not likely, as rather a lot of them were full of ale and high spirits by then!

In Exeter, Devon, another surviving relic, a large stuffed glove decorated with ribbons and garlands, is still paraded through the town and then displayed atop the Guildhall roof. The town crier calls out:

> "Oyez! Oyez! Oyez!
> The fair's begun, the glove is up.
> No man can be arrested until the glove is taken down."

William Shakespeare's father, John, was a glover who operated from a shop adjoining the family home in Stratford-on-Avon. It has been suggested that William may have helped out there as a youngster; certainly he makes many references to gloves in his plays, such as "Scented gloves as sweet as damask roses," in *The Winter's Tale*, and Romeo's speech wishing to be a glove so that he might touch Juliet's cheek. Shakespeare also has scenes in Henry V, written in 1598, where the King in disguise moves through his army and is challenged to exchange gloves and fight next day. The poor man is horrified when he discovers that he has inadvertently challenged his King, but is pardoned for what would otherwise be treason, punishable by death.

TO TREAT SOMEONE WITH KID GLOVES or WITH VELVET GLOVES means **very gently**, as gloves made of the skin of a young goat were especially fine and delicate and suggested a refined person who you would **treat with care**.

When we speak of someone as having AN IRON HAND IN A VELVET GLOVE we mean they are **kind and fair, but with underlying steel** in their personality. The phrase seems to have been originated by Napoleon and popularised by Thomas Carlyle in the mid-19th century.

An example of such firmness occurs in a mouse-sized story from the Welsh epic tales of the Mabinogion. The Lord of Dyfed is beset by his enemy, Llwyd the Enchanter, who has kidnapped his wife Rhiannon and his stepson Pryderi and has sent a plague of mice to eat his seed corn. The Lord catches a female mouse in his glove and insists he is going to hang her for a thief. It so happens that the mouse is Llwyd's own wife, shape-changed by his magic. The Enchanter then assumes many different shapes himself, trying to outwit his opponent. He also tries to ransom his wife; but in vain. Finally Llwyd is forced to reveal himself in his true form. The Lord of Dyfed makes him promise to return his captives, do no more harm to the country, and never take revenge. Only then does he release Dyfed's mouse-wife, after which, his beloved Rhiannon and Pryderi are returned to him safely.

Superstition says that a dropped glove should never be picked up by the person who dropped it. "Oh dear, I've dropped my glove. It's bad luck to pick it up myself." you might say, fluttering your eyelashes appealingly at a passing man. With luck, his gallantry could result in – who knows? Dropping your glove, hankie or fan was a well-known ploy to broach acquaintance in Victorian times.

ON ONE HAND OR THE OTHER, we say when contrasting two things; and what one actually wears on one's hands can convey widely differing messages.

For example, a clean white glove stands for innocence. A pair was traditionally given to the Assize Judge in England if there were no criminal cases to bring before him.

White gloves denoting purity also arise in the traditional funeral practice that marks the death of a young girl or spinster. The undertaker wears white gloves, and girls or fellow spinsters in white gloves may place ropes of flowers known as Maiden Garlands on the grave both at the burial service and each year thereafter on St Faith's day, October the 6th.

However, gloves can be used for evil purposes. In 1619, Joan Flower and her two daughters were found guilty of killing the young Lord Ross, heir to the Earl of Rutland, by witchcraft. They were accused of stealing one of his gloves and rubbing it against their 'familiar', or demonic assistant, a cat called Rutterkin. Supposedly, they then dipped said glove in water, pricked it, and buried it in the earth. Soon after, the Earl's whole family fell ill with terrible convulsions and the boy, his heir, perished. Following what has become known as the "Witches of Belvoir Trials", the two Flower daughters were hanged at Lincoln. Their mother refused to confess and asked for bread and butter; she said "May I choke if I am not innocent." Immediately she took her first bite, she choked and died. What a grim little story!

The monster Grendel from the Old English epic poem Beowulf was said to have a *glof* or glove, usually taken to mean a bag. It was huge. So was that of the giant Skrymir. Indeed, it was so big that Thor and his companions spent the night inside it, thinking it was a great hall.

Anything that FITS LIKE A GLOVE **fits perfectly**, while GLOVE MONEY – a little something that can be insinuated into the palm of a glove is **a bribe**, cash passed over secretly. Or it may be simply a tip.

The FOXGLOVE plant had nothing to do with the animal, but referred to its sacklike flower as a GLOVE said to belong to **the folk**, which is a euphemism for **the fairies**. Be careful always to use such oblique words to refer to these magical beings, as you may arouse their enmity if you speak of them too directly.

GRAY EMINENCE ☞ CAPPUCCINO

A GRAY EMINENCE or *EMINENCE GRISE* if you want to use the original French, denotes **a shadowy figure pulling the strings** of an enterprise, but never coming into the light to be held accountable. Capuchin monks were known as GRAY MONKS because of the more-or-less buff shade of their robes. Either the dye or the eyes of the beholder must have varied, because gray-coloured coffee doesn't really appeal.

One of these monks, known as Père Joseph, was widely believed to be an evil influence on Cardinal Richelieu in early 17th century France. A book about him entitled *Grey Eminence*, written by Aldous Huxley in 1941, brought the phrase into popular use.

HABERDASHERY ☞ PICCADILLY

This word is a favourite of mine, because of its sound – like softly swishing satin skirts. It is a term used for a mixture of all sorts of useful little sewing essentials and knick-knacks.

In modern urban slang, haberdashery is now a noun referring to **foolish acts verging on hooliganism** by drunken young men (or women) in public places. You might say, "Those lads in the town centre last night were up to all sorts of tomfoolery! Having a right HABERDASHERY, they were! Drunk, of course!" It embodies a certain note of tolerant amusement, and can be found via the Internet in the Urban Dictionary.

I think we should claim the word back! Many of us stitchers have been known to commit acts of HABERDASHERY, though without the vandalism and

··· *Haberdashery* ···

alcohol. Almost every time we are exposed to stalls or shops selling thread, ribbons, buttons, collars, zips, garters, needles and pins, laces, hats, collars, and unidentifiable 'sewing notions' we are transformed into glassy-eyed zombies, impervious to reason. We suddenly feel we cannot live without them, even though we haven't a clue why, or when we will have the time to use them.

"It was just an attack of HABERDASHERY. I'm over it now!" we may cry. Our friends will start gossiping, "She's having treatment for her tendency to HABERDASHERY, you know. I recommended she join HABERDASHERS ANONYMOUS, but she still can't give up her stash."

So, where did this word come from? The [sober] Shorter Oxford English Dictionary says that it started as *hapertas,* of unknown origin and meaning. By the 15th century, HABERDASH meant 'small wares', often caps and hats in lace and delicate fabrics. By the 17th Century, the term was also used for the trade in collars, cuffs, belts, fastenings, threads, ribbons and tapes – the sort of thing a travelling peddler might carry from village to village. They were dry goods and textile-connected. Indeed, the terms Haberdasher, Mercer, Hatter and Draper, were sometimes used interchangeably, though they more properly distinguished between small wares, silks, hats and furnishing fabrics respectively.

In America, the term migrated to shops selling **men's wear**, especially cheap suits, the sort of emporium where an invading tribe of haberdashery-crazed women might provoke rather a stir.

Back in the UK, the Haberdashers' Guild was granted a Royal Charter in 1448, and now is known rather grandly now as "The Master and Four Wardens of the Fraternity of the Art or Mystery of Haberdashers in the City of London." They are a gracious and dignified body, given to extensive charitable work, in particular the founding of Haberdasher Schools all over the UK. I wonder if the lively minds of their students are responsible for the modern slang word HABERDASHERY as a term for **high-spirited mayhem**?

HABIT

The word HABIT came originally the Latin *habere* signifying 'to have', and referred not only to clothing but also to appearance and demeanour. It then moved on from being **something one has or wears** to its modern meaning of **something one regularly does**. At the same time, it kept its clothing link for one specialised meaning: a garment or style of dress one normally **wears**, rather like a uniform. These days, while all of us still have lots of different kinds of behavioural habits – good, bad and decidedly ugly – only monks, nuns and horse-riders still wear HABITS.

Incidentally, another word used for the outfit you customarily wear is COSTUME, and when we speak of someone 'wearing a costume', either in the sense of theatrical dress or a woman's suit, we are actually talking about 'a custom', from the Latin *consuetude*. Both the French and the Italians used the idea of customary things to refer to a general style belonging to a certain period; this could refer to artistic creations, furnishings and decor, as well as clothes. It was only in 19th century England that COSTUME was restricted to the **clothing seen in any particular period**, and hence, to **theatrical clothing**.

HANDBAG ☞ BAG

Pretty obviously, a handbag is meant to be 'a bag you can hold or carry in your hand'.

"I'LL HIT YOU WITH MY HANDBAG!" is used as a camp sort of threat, as of **a light blow from a gay man**, and it has been contracted to the very modern HANDBAG to describe **a minor spat** or hissy fit in sport. "IT'S HANDBAGS AT DAWN!" someone might say, over-dramatically, in tribute to more serious duels from earlier times.

Some little pet dogs are specially bred to to be portable, and can be carried around like any other fashion accessory. They are sometimes popped inside a chic *sac à main* and therefore known as HANDBAG DOGS. Goodness, I hope they haven't been given too much to drink; things could get a bit messy.

Britain's formidable former Prime Minister, Margaret Thatcher (later made a Baroness), was said **to intimidate** or HANDBAG her male Cabinet members, browbeating them into submission to her will. And they loved her for it (well, some did).

HANDKERCHIEF ☞ BANDANA and SLEEVES

I really would have preferred to include HANDKERCHIEF in my entry on sleeves because, from time immemorial, men's shirtsleeves have been doing duty to mop up spilt beer, wipe noses, and dry other unmentionable places. Ladies, of course, were meant to possess delicate cambric squares surrounded by lace, of no practical use. After all, the delicate sex were not supposed to have anything as vulgar as the common cold or a sweaty forehead; a saying from Victorian times went, "Horses sweat, Men perspire, but Ladies only glow!

The handkerchief effectively became obsolete with the advent of disposable paper tissues. If these had been invented sooner, they would have robbed us of so many cultural and literary associations. These traditional, small squares of fabric are full of significance in dramatic plots. One has only to think of Othello in Shakespeare's eponymous play, where the villain Iago plants Desdemona's handkerchief in a compromising place, misleading Othello into believing that his wife has been unfaithful to him. The Moor's jealousy maddens him and he chokes his beloved to death. A disposable tissue would have shortened this play by 99%, and ruined the whole plot.

So, although it is only a small scrap of textile, the handkerchief wraps up a hefty package of folklore, magic and fun! The word is derived from the obsolete 'kerchief' meaning a 'head covering'. In Old French it was *couvrechief*, combining *couvrer*, 'to cover', with *chief*, 'the head', later contracted to *courchef*. It therefore started life as a simple scarf.

You might be instructed TO TIE A KNOT IN YOUR HANDKERCHIEF in order **to remind yourself of something**. That's definitely not easy to do with a shirtsleeve, especially while your arm is still in it, so the invention of the hankie was a Very Good Thing. Curiously, no one has ever been recorded as remembering what their hankie knot was for.

To stimulate your brain (if not your memory), you might enjoy an old-fashioned tongue twister:

> "My grandmother sent me a new fashioned three-cornered cambric country-cut handkerchief,
> Not an old-fashioned three-cornered cambric country-cut handkerchief,
> But a new-fashioned three-cornered country-cut cambric handkerchief."

The HANKIE or HANKY is, of course, an abbreviation of handkerchief and sprang up in the 19th century. HANKY-PANKY, is one of our lovely, rhyming double words. It suggests something **tricksy** or **underhand** and especially a bit of **sexual messing around**, such as a brief affair that nobody took very seriously. This term originated with the legerdemain of a conjuror who hid his flim-flammery behind a showy silk handkerchief.

There are many traditional stories regarding the magical properties of fabrics — even the small and humble handkerchief.

Saint David, patron saint of Wales, was enormously popular and well-loved. It is said that when he preached, so many people congregated that most were unable to see him. For one jam-packed gathering, rather than disappoint the crowds, he stood on his handkerchief and — lo and behold! — the hankie levitated with him on it while he preached, so that he was comfortably visible to all. Once he was done, the hankie returned him safely to the ground. He was definitely a "man of the cloth"!

My favourite handkerchief story brings together Queen Elizabeth I of England and the Irish Pirate Queen, Grace O'Malley. Grace's name was Granuaile Ui Maille in Gaelic. This 16th Century Irish clan leader and sea trader fought many sea battles both against other clans and the English. Queen Elizabeth I became very interested in her – no doubt owing to her depredation of English ships – and in 1593, a meeting took place at Greenwich between these two formidable women. The story goes that Queen Elizabeth was pleased to give Grace a ceremonial HANDKERCHIEF, embroidered with pearls and gold thread, a great sign of condescension. Grace O'Malley blew her nose on the Royal hankie and dropped it into the fire. The Queen, eyes flashing, drew herself up and said "The line I come from would have sent someone to the Tower for lesser insults than that!" to which Grace replied, "The line I come from does not need to use a handkerchief more than once." After a long pause the two ladies parted, heads held high, honours being even.

Do we catch our snot or bogies in a handkerchief? [I hope so!] Here's a story of how a Bogey nearly caught the man instead!

Once upon a time, a man called Callum was going home late from market when he realised he had lost his pocket handkerchief; so he retraced his steps. It was well after midnight when he came to a stream and saw a Bogart or Bogey Man by the waterside, washing and rubbing the very same HANKIE on the stones. They fought long and hard for that scrap of cloth, but eventually, Callum wrested it away. As he departed victorious, the Bogart shouted after him, "It was lucky you came back. I'd have rubbed a hole in it by morning, and that would have been your death!" So be careful where you chuck your used paper tissues. It wouldn't take two seconds for the Bogey Man to rub a hole in one of those.

Any personal possession in folklore carries something of its owner's essence and therefore can be used magically for healing or harm. That may add an underlying resonance to the plot of Othello. As for the nose-wiping type of hankie, a repository of bodily secretion, it would be particularly effective in magic terms – and a good source of DNA in a criminal investigation!

I expect that Callum possessed a true, old-fashioned linen HANDKERCHIEF. Linen has remarkable strength, resists wear by rubbing and becomes more and more lustrous and tough each time it is washed.

Such large, old-fashioned hankies, with knots in their corners, are still occasionally seen on the beaches of Britain as sunhats for men, though this has become a rare and treasured cultural anachronism – even among older Brits. Traditionally, a knotted hankie was always worn with a full suit – probably to indicate that the wearer had come in his 'Sunday best', or only just arrived, hot-foot from his office and should therefore be Taken Seriously. It was considered permissible to remove the jacket to sit on, and the bottoms of the trousers could be rolled up against the shins as a concession to beachwear. Only the British would understand or indulge in this eccentric use of a HANKIE.

Other traditionalists wear a neatly folded linen or silk HANDKERCHIEF – never to be used – in the top left hand pocket of their suit jacket. This is a last, brave vestige of 18th century underlinen and the shirt or colourful camisia peeping through the slashes of a Renaissance doublet. It's a [silent] sartorial figure of speech intended to say, "I may look like a boring 21st century twit, but beneath this cloth beats the blood of a cavalier man-about-town." Judging by the cut and thrust of City life, I reckon they are getting in touch with their inner RUFFIAN.

HAT ☞ CAPE, HOOD

Both the hat and the hood come from an Indo-European root thought to be *kadh* meaning 'to cover' or 'protect', leading through the Germanic language *khadnus* and *khattus* to become A HAT. The Dutch call such head covering a *hoed* which is nearer to our 'hood', and the word is linked closely to *cappellus*, meaning 'a hooded cloak'.

To tell someone I TAKE OFF MY HAT TO YOU means that **you admire them** very much. It is a purely metaphorical gesture in these hatless days; however, in ceremonies, such as in venerable universities or before royalty, the doffing of headgear may still be seen. Occasionally, delightfully genteel, old fashioned men still TIP THEIR HAT to a lady – at least figuratively – by treating her politely. A HIGH HAT is **a proud person**, and an early 20th century North American gent who was given the brush-off or treated superciliously by a lady might say, **"She high-hatted me"**. To say something is OLD HAT is to dismiss it as **out of fashion** or just a bit tired and passé.

The phrase for **keeping something secret**, TO KEEP IT UNDER YOUR HAT has been known since the 19th century, but it only became popular in the 1940's, when Cecily Courtneidge and Jack Hulbert appeared in London's West End in a musical called "Under Your Hat". In World War II, it was employed in government propaganda against idle talk, appearing on posters as KEEP IT UNDER YOUR HAT; CARELESS TALK COSTS LIVES. Since then, the phrase has become something of a cliché – or OLD HAT!

In Celtic history, one man who certainly had to KEEP IT UNDER HIS HAT was the hero Diarmaid, who was cursed (or blessed) with a curious birthmark on his forehead that caused any woman seeing it to fall passionately in love with him. He had to wear a hat all the time to avoid being weighed down by clinging females. However, this is not an excuse that your average spotty teenager in a baseball cap should try – especially when he sits down at table to eat with his family. Take that hat off, young man!

'IF YOU WANT TO GET AHEAD, GET A HAT', was a slogan dreamed up by the British Hat Council, in the early 1950's, but perhaps known earlier. It means that **you will succeed better in life if you wear a hat**. Supposedly, a hat will make you appear more serious, wiser, richer, more elegant, and altogether a Better Person. Now, why do I suspect that they said it simply to sell more hats?

If you are TALKING THROUGH YOUR HAT you are making no sense, a phrase which seems to have started in the USA in the 19th century and meant originally to **bluster**, perhaps becoming **to talk nonsense** because someone who is angry splutters and gets their words mixed up. Another term from America is ALL HAT AND NO CATTLE; originally it meant braggarts who wore the right cowboy clothes but had no ranch, no livestock, and probably couldn't even ride a horse. It later meant anyone who talks big but hasn't got the skills or substance, just as with **all mouth and trousers**.

To call someone a BAD HAT means you think that he is a really **awful person**. It may have started out as A SHOCKING BAD HAT, which is of uncertain origin, but was certainly current in the 19th century. It might have come from the hats worn by ancient Irish kings or even have been coined by the Duke of Wellington. Maybe you would feel justified with such a person to KNOCK HIM INTO A COCKED HAT, meaning, of course, hit him very hard and repeatedly **till he was all folded up** like a piece of paper into the shape of an 18th Century tricorn [three cornered] hat.

I'LL EAT MY HAT IF... is **a strong statement of belief**, and first found in print in Charles Dickens' *Pickwick Papers* in the 19th century. However, it may well have been known earlier as it was quite common when asserting a point with vigour to vow, "I'LL BET MY HAT ON IT,." as a similar strong statement of belief.

To say of someone HIS HAT COVERS HIS FAMILY is an old fashioned and rather sad way of saying **he is alone in the world with no relatives**, while TO HANG UP YOUR HAT could mean **to get engaged**, or to **make yourself at home** or even **to die**, or even, in one way or another, to "settle down" – possibly for eternity. Today, we use it more commonly to mean **retiring**

from work. A happier phrase is HOME IS WHERE YOU HANG YOUR HAT, suggesting **you can have peace and contentment wherever you are, if you have the right attitude**. There is indeed a song with that title, written by Marvin Gaye, Barrett Strong and Norman Whitfield, and first recorded by Gaye in 1962.

It has long been a common practice for street performers such as buskers or the old organ grinder and his monkey to pass a hat around the audience to get a few coins. Well, the man might expect hard cash, but presumably the monkey would work for peanuts. If you are due to retire your work colleagues might SEND ROUND THE HAT **to raise money to buy a present** for you, particularly if you **had two jobs** and were described as WEARING TWO HATS. This last is probably recent and was apparently first recorded in The Times in 1963.

You might PULL A RABBIT OUT OF THE HAT if you were a conjuror, or use it metaphorically for **pulling off any sort of coup**. It's a comparatively modern term dating from the 1960's. The very similar metaphor TO PULL OFF A HAT TRICK is a sporting figure of speech used to illustrate **getting three of anything in a row**, particularly goals in football, or figuratively, **to do something unexpected and almost magical**. It came originally from the cricketing practice of awarding a team cap to any bowler who could dismiss three batsmen with consecutive balls.

"WHERE DID YOU GET THAT HAT, WHERE DID YOU GET THAT TILE?" came from a popular music hall song and you might sing it to anyone **to show you think they look a little silly**, while the term **titfer** comes from 19th century Cockney rhyming slang – 'tit for tat' equates to HAT.

The origin of the term MAD AS A HATTER is subject to many words of debate, but the likeliest explanation is that hatmakers of old were exposed to mercury vapours when doing felting, and the poisonous fumes produced symptoms such as 'hatter's shakes' with twitching and tremor, irritability, memory loss, depression and even frank psychotic episodes. Indeed, there is a medical condition known as MAD HATTERS' DISEASE. The phrase is commonly used in the UK nowadays to mean **extremely eccentric or crazy**, and Lewis Carroll popularised it with his Mad Hatter character in Alice in Wonderland, written in 1865.

Alternatively, it has been suggested that the term was originally **mad** as in angry, being mad as an adder, or an otter, or even as an attercop, which last is an old name for a spider and means 'poison head'. The term was certainly also used in the USA from the early 19th century, with **mad** in the sense of **angry** rather than **crazy**.

To say MY HAT IS IN THE RING means that I am **looking for a fight** because travelling prizefighters would go round the fairs challenging the locals to fight them. The men would throw their hats into the ring and each hat in turn would be held up for the owner to claim, or he might spontaneously –

"Throw in his hat, and with a spring
Get gallantly within the ring."

Political candidates often say I'M THROWING MY HAT IN THE RING, to mean they have officially started electioneering and want your vote.

In the Wild West, the start of a fight or a horse race was signalled by a man dropping a hat. Therefore, AT THE DROP OF A HAT came to mean **anything that happens very quickly**.

As so often happens an item of clothing can come to signify a whole group of people. A BRASS HAT was originally an Army Officer of high rank **wearing a lot of gold braid on his hat** and the term is now used as THE TOP BRASS or THE BIG BRASS for any dignitary or boss. It is not really used as a term of respect; like most of these phrases reducing people to one thing, and that their outward appearance, carries a hint of contempt. Another theory of the term's origin is that BRASS CAP referred to the cocked hat worn by Napoleon and his officers, which they folded and carried under their arm when indoors. In French these were called *'chapeaux à bras'* ('hats in arms'), a term that the British anglicized to 'brass.'

A BOWLER HAT, BRIEFCASE and UMBRELLA were the marks of a city gent. Therefore, the phrase THE BOWLER HAT BRIGADE is uttered a little scornfully of **a group of bureaucrats or civil servants** whose working attire was originally formal, suggesting **someone rigid and culture-bound**. The original Bowlers were a 19th century family of London Hatters who developed the characteristic pudding shaped felt hat with a curled brim and gave it their name.

The term BOWLER HAT can be used on its own to denote a clerical, non-manual worker. In the somewhat sleazy Bowery neighbourhood of late 19th century New York, a group or gang of young toughs and hard drinkers were known as ROUND RIMMERS from the **brimmed** BOWLER HATS they wore.

HOOD ☞ COAT, CLOAK, HAT and VEIL

A HOOD is a piece of cloth or other fabric that partially covers a person's head, leaving a gap for them to see out of. It may be separate from the other garments, or attached to a COAT or CAPE. The hood often protects both hair and face, but can also be used to disguise the person's exact facial features and confer anonymity, so it has a mixed reception among onlookers.

The idea that you may be **concealing sinister intentions** behind a HOOD has led to people being banned from wearing them in many Town Centres, where in recent years young thugs have sometimes rampaged through the streets with the hoods of their fleecy jackets up to avoid identification. Referred to as HOODIES, these hooligans have made life difficult for harmless elderly joggers who find themselves unfairly banned from shopping malls.

The slang HOOD is also a shortened form of 'Neighbourhood,' a term often used by American gangs to **mark out the area** they claim. Ghetto slang today uses HOODS to mean **gang members** from their own neighbourhood, who are also called 'homies', as they are from the home turf. But A HOOD meant a 'bad boy' more than half a century ago. Certainly as long ago as the 1950's it was slang meaning **hoodlum, a petty criminal or teenage delinquent**. But that's not the end of it! Travel back to 1868, to the streets of San Francisco, and you might have been set upon, roughed up and robbed by the no-good Hoodlum Gang. Their name may have been a corruption of the Bavarian dialect term *Huddellump* or ragamuffin. And Voila! We are back at another textile connection.

We speak metaphorically of **a mountain covered** in a HOOD OF MIST or OF CLOUD or as generally HOODED. We can also use the term figuratively to suggest things that are **partly concealed** such as a man's HOODED EYES, where you cannot see what he is thinking, as **the lids are half closed**.

TO HOODWINK SOMEONE is to fool or trick them by not showing your real intentions. This phrase first used literally as **to cover someone's eyes with a blindfold**, and its metaphorical use of **to cheat or deceive** was first recorded in the 17th century.

In the US, the **hinged lid** over the engine of a car is called a HOOD, whereas in the UK it is a BONNET [qv]. Either way, clothing-related ideas are inescapable.

HOSE

The word HOSE originally meant 'stockinged feet'. It came from a vanished Germanic word *khuson* for a leg-covering, which led to the German and English HOSE. In the 15th century we adopted the use of HOSE as **a long tube for conveying liquid** from the Dutch *hoos*, where it was already in use metaphorically for this purpose.

An Old English song, the Lyke Wake Dirge, is about the soul approaching death, represented by crossing a wild place, Whinny Muir [or Moor] and contains the chilling lines:

> If ever ye gaven Hosen and Shoon [stockings and shoes]
> Sit ye down put them on.
> If Hosen and Shoon thou ne'er gave none,
> The thorns shall prick ye to the bare bane.

INVEST YOUR MONEY ☞ DRAG, TRAVESTY and VEST

The early 17th century Italian verb *investire* – the prefix *in* plus *vestis* denoting garments – is thought to have drawn on the idea of **dressing up one's capital in different clothes** by varying the business or stock one purchased with spare cash. Hence, we use INVEST to mean **putting money into a financial scheme or shares**. It gives me a slightly warmer feeling for all those City financiers, speculating away and making millions when I realise that they are metaphorically playing at dressing up. I love the idea of a cross-dressing investment banker with a frilly dress, bowler hat, rolled umbrella, moustache and hairy legs.

Once upon a time, important men of business in London wore very fancy garments. As the Nursery Rhyme says,

> "Hey diddle dinketty, poppety, pet,
> The merchants of London they wear scarlet;
> Silk in the collar and gold in the hem,
> So merrily march the merchantmen."

This little jingle probably originated in medieval times. It seems to celebrate the various Guilds of Merchants of the City of London that came into being in the late 12th century. The City's business centre took up about one square mile, which today is still packed with streets commemorating these trades. The Guild members – masters, journeymen and apprentices – dressed in the distinctive clothing of their various occupations: mercers, goldsmiths, grocers, etc, for processions on saints' days. Their garb, often quite sumptuous and created in rich fabrics, gold lace, fur and so on, was known as livery, because this originally meant a uniform given to servants from their masters. Eventually, the guilds and trades became proper corporations known as Livery Companies, and collectively, they are simply called 'the Livery'.

Staying with the same formal, ritual theme, we move to the idea of an INVESTITURE, which also comes from Latin *investire*. For several centuries, it kept its literal sense of **clothing someone**; but today, being invested is only used metaphorically to mean **being given official recognition**, enrolled in some group, or **installed in a high position**. Even so, it is still sometimes still marked out by the wearing or giving of special ceremonial robes. Royalty

··· Invest your money ···

and some academics are still literally clothed or invested while wearing traditional garb and carrying tokens of their rights or powers. An INVESTITURE is also the ceremony when private individuals receive honours or medals recognising merit directly from a member of the Royal Family. Charles, heir to the British throne, was INVESTED as the 21st Prince of Wales by his mum, the Queen, back in 1969 at Caernarfon Castle, and received the insignia of his Principality and the Earldom of Chester: a sword, coronet, mantle, gold ring, and gold rod.

The term TRANSVESTITE for someone who **cross-dresses** was from the same source, but only came into common use in the early 20th century via German *Transvestit*, although the very similar idea, *travestiment*, meaning **to wear the clothing of the opposite sex** was recorded in Britain back in 1832.

JACKET ☞ CLOAK, COAT

The JACKET came from the Old French generic name for a man – *Jacques*. It essentially denoted **a peasant** and the garment in question was close fitting, like chain mail. Some sources suggest the word derived from an Arabic breastplate known as a *shakle*. Both these stories, of course, could be true, since a word that sounded a bit like '*Jacques*' to a rough local soldier might have become a nickname for the garment. It came to England as a stuffed or quilted protector for the chest, and by the 14th century, was more or less amalgamated into the COTE.

That French "peasant" nickname has also given us the term DUST JACKET for the **protective cover over newly published hardcover books**, first noted in the late 19th century. And we might place a CYLINDER JACKET, meaning **an insulating or protective layer** on a hot water immersion heater, or a shrapnel-proof FLAK JACKET on a soldier – which brings us neatly back to a modern-day 'Jacques'.

Also, as we all know, we use the term a JACKET POTATO for those lovely fluffy spuds with their protective skins on, baked in the oven or bonfire and filled with butter and other delicious things. Yum, yum, yum.

Speakers of American English use the terms **jacket** and **coat** almost interchangeably, and most of us would be hard put to it to describe the difference, except that typical modern jackets extend only to the upper thigh in length, whereas a coat is normally a bit longer, and usually of heavier fabric.

Older-style coats, such as tailcoats, usually reach the knees. The modern jacket worn with a suit is traditionally called a lounge coat (or a lounge jacket) in British English; it used to be known as a sack coat in American English, but now, this term is rarely used. Most people in the US today would simply say a 'suit jacket' for a matching part of a formal outfit, or 'sports jacket' for the more casual style that may be of different material from the trousers or "pants".

Because the basic pattern for the 'stroller' (a black jacket worn with pin-striped trousers in Britain) and the dinner jacket (tuxedo in American English) are basically the same as for lounge coats, tailors traditionally call both of these special types of jackets a **coatum**.

Before the 1960s and the sartorial "whatever's right, man" attitude this decade ushered in, the majority of men dressed in a coat and tie; now we would say "jacket and tie". Either way, such formality is rarer, except for managers and professionals at the office.

An OVERCOAT is designed to be the outermost garment for outdoor wear; while this word is still used in some places – particularly in Britain – elsewhere, the term COAT is the common shorthand way of saying the same thing. A topcoat is a slightly shorter overcoat, if any distinction is to be made; this is what Yanks call a CAR COAT. In the early, breezy days of motoring in open cars that often lacked heaters, drivers and passengers bundled up in long, cover-all fur coats or lighter weight 'dusters'. Over time, as cars became cosier, these coats got shorter, and today's version normally reaches mid-thigh.

Overcoats worn over the top of knee-length under-coats including frock coats, dress coats, and morning coats, are cut just a little longer so as to completely cover the lower layer, as well as being roomy enough to accommodate everything worn underneath. Confused? Don't worry, aren't we all! You really don't need to worry about any of this unless you are a tailor!

JUMPER ☞ LOOSE WOMEN, PULLOVER and SWEATER

A JUMPER in Britain is a sleeved, knitted item of clothing for the upper body, easy to pull on and off – a casual word for a casual garment that North Americans know as a **sweater**. The word came from a dialect word *jump*, allied to the *jup*, which was a short coat for men or a sort of woman's underbodice. It is linked to the French *jupe*, meaning a skirt, and perhaps much earlier, derived from the Arabic *jubbah*, which is a loose ankle-length robe similar to a kaftan.

In the 18th century, the LOOSE JUMPS worn by women were a much more comfortable and relaxed form of corsetry than the earlier underwear stiffened with busks of whalebone or wood. (see LOOSE WOMEN)

From the French, who gave us *jup*, we also derived the dialect word 'jump'. That not only led to the easygoing undergarments called JUMPS, but also to the JUMPER, a knitted, cardigan-style or pullover top. It's another cosy, comfort garment. In the 1940's and 50's especially, men talked about their favourite SWEATER GIRL, who was sometimes a movie star pin-up, but could also be any **young woman with a big bust**, outlined by clinging wool.

OOMPAH, OOMPAH, STICK IT UP YOUR JUMPER [pronounced 'joompah' to rhyme] is **a dismissive phrase, rather vulgar**, possibly originating in a music hall song. Oompah or Oompa-pa-pa has been used to imitate the fart-like sounds that can be made by brass instruments since at least the 1920's. Like other naughty 'stick it up yours' phrases, accompanied by rude gestures, this taunt is generally considered good-humoured enough not to cause any real offence.

KNICKERS TO YOU TOO! ☞ PANTS

It's time we got down and dirty to look at a few of the figures of speech derived from women's underwear. When we consider these intimate garments, we are touching on areas [no sniggers at the back, please] regarding clothing that historically evoked powerful fascination and shame in equal proportions. But don't worry; I only intend to look at the more respectable phrases and figures of speech that you might hear everyday. [Do I hear you cry, "What a pity!"?]

By the 1920's, elastic had become fairly reliable for holding up stockings, and flirtatious girls really wanted their lacy or brightly-coloured garters and suspenders to be seen. The really daring young things might even go without knickers, so that climbing up into the rumble seat of a car offered very much the same flash of delight for the boyfriend as the high-flying garden swing occasionally did for the 18th century beau.

"OH KNICKERS!" is used as a **very mild, almost harmless expression of exasperation**. Both this and "KNICKERS TO YOU!" as another **mild curse**, could be regarded as slightly twee in use, now that some women make free with all the rude and shocking language they can.

It's a little odd to realise that, until about 200 years ago, it was the wearing of garments round the lower torso, not the absence of KNICKERS on women that was seen as lewd and morally depraved. 'Drawers' were only for men until the late 18th century, when the gauzy, French Empire-line dresses proved so chilly and translucent that ladies' pantaloons [see PANTS] reaching down to the knee became *de rigeur*.

Earlier, KNICKERS was a plural word, short for 'knickerbockers'. These were men's loose-fitting, knee length trousers gathered in below the knees, and named after a fictional early New York settler invented by the American 19th century author Washington Irving – who, by the way, also concocted Rip Van Winkle and *The Legend of Sleepy Hollow* with its "headless horseman". The writer thought "Diedrich Knickerbocker" sounded suitably Dutch, and his creation was depicted wearing knickerbockers in Cruickshank's illustrations for Irving's *History of New York*.

These days we might also exclaim, DON'T GET YOUR KNICKERS IN A TWIST meaning **keep calm, and don't get uptight about the situation.**

LAUNDRY LIST

Originally, A LAUNDRY LIST was obviously what a commercial laundry sent home with your washing to account for the odd socks, and tattered rags that had once been your perfectly-matched and intact, precious clothing. The word LAUNDRY simple comes from roots to do with **washing clothes**, but became a term for a collection of miscellaneous items ready to be cleansed or the place in which that could be done. A LAUNDRY LIST was later applied figuratively to **any list of anything**, but tends to refer to endless, tedious tasks you would really rather not do.

A similar metaphoric use, with words about washing only implied, appears in TO HANG YOU OUT TO DRY, said to an enemy or someone you intend to punish. This is certainly not a nice threat, but it doesn't mean to punish you by hanging until you are dead. It means to HANG YOU OUT LIKE LAUNDRY, sometimes with the addition of LEAVE YOU BLOWING IN THE WIND – in other words, **to expose you to the full light of everyone's gaze**, like washing on a line. Such a threat **to make it a public affair** – reported in the Press or on Twitter day after day, might well make you feel that you are being treated like laundry.

The idea of dirty clothing is very much present in scandal, however well the subjects are dressed. Indeed, the phrase WASHING YOUR DIRTY LINEN, or AIRING YOUR DIRTY LAUNDRY IN PUBLIC means just that – **exposing shameful secrets** like mucky underwear seen and despised by everyone. How embarrassing!

LEATHER ☞ FILLET

Articles made of skin, hide or leather can be used for many, many things – for sandals, gloves, bags, hats, clothing, sails, bedding, parchment or vellum pages to write on, and little boats called coracles, where skin is stretched over a bendy, wooden framework and then smeared with pitch or tar. In the past, leather straps made factory or machine gears turn, and I.K. Brunel used leather flaps in his innovative 'atmospheric railway'. Animal skin can be used for just about anything that CLOTH can, with the added benefit of being stronger, and often waterproof.

Since very ancient prehistory, humans have used the skin, hide or pelt of creatures to make a strong form of fabric. A hide had to go through long processes of being soaked or buried in bogs, scraped, tanned by natural chemicals and generally messed around with by expert leather workers and tanners. The result offered an important, versatile raw material employed by all civilisations, including the Romans, who used it for armour, belts, buckets and tents. They developed and codified ways of treating the skin in order to make it durable, supple and – hopefully – less smelly as the organic matter eventually rotted.

As a surname or place name anything with LEATHER or TANNER in it is likely to have had such workers in their pedigree. In London, one of the earliest of the famous guilds is the Worshipful Company of Leather Workers. Some such names, however have been shown to come from a different source altogether; a rather similar Old Norse word, *hleother*, meaning 'sound' or 'melody'. So I like to think that the Leatherworkers Guild can sing particularly tunefully, as indeed Cobblers sometimes do at their work.

To say you are LEATHERING SOMEONE or GIVING THEM A LEATHERING is a way of saying you are **giving them a severe beating** – as with a leather whip. In the same way, you can describe someone as GOING HELL FOR LEATHER, meaning **extremely fast**, from beating your horse with the ends of the reins or a whip to make it work harder. It was sometimes HELL BENT FOR LEATHER, to emphasise **the determined nature** of your ride; in both, the expletive 'hell' is used simply as an intensifier. The phrase is still used nowadays, even though whipping your poor old car as John Cleese famously did in the TV show Fawlty Towers isn't likely to do anything other than getting yourself looked at rather oddly.

A similar phrase is TO SLIPPER someone. It implies a beating or smacking that is supposedly a little gentler and was used on children, using a leather slipper rather than a whip. Those truly were the bad old days.

In the late 18th century, A LEATHERNECK was first used in a derogatory way for **soldiers of the USA's Marine Corps and the British Royal Marines**. The slang term is said to be derived from the use of a heavy LEATHER COLLAR OR STOCK, reinforced with metal, to be worn at all times –

indeed, even sometimes stitched to the neck of the uniform. This served the double purpose of keeping the soldier's head high, no matter how tired or ill he felt, and also protecting him from the slash of a sword or cutlass. By the way, US Marines are still called LEATHERNECKS, and are proud of the epithet that began as an insult.

LIMOUSINE

If you are lucky enough to be taken here and there in a luxurious vehicle driven by a chauffeur and call the car A LIMOUSINE, please be aware that you are actually running round in a special caped cloak. Originally this article of dress, a long cape worn by shepherds, took its name from the Province of Limousin in France. Eventually, a LIMOUSINE was the nickname given to **a motor car in which passengers were enclosed within a private compartment, with a driver sitting behind sliding glass panels under a projecting roof** rather like the garment's hood.

These vintage cars have become popular in recent years in America, and there is even a National Limousine Association in the UK. The Guinness book of world records describes a car of 30 metres long, complete with swimming pool, helipad and putting green. Try parking *that* outside the average suburban home.

LINING

I really have to mention LINING as a relevant textile word, although it has been extensively covered in **Rigmaroles and Ragamuffins**, with some 70 uses of LINE as a composite word, phrase or metaphor!

It came originally from the Latin, *linum*, for a thread of flax and the Latin *linea* for LINE and is extensively used with this meaning, and also for a piece of fabric, or any other substance, used alongside something else as a LINING. In the 15th century it was spelt, rather endearingly, LYNYNG, which would be a good word for one of those tricky games of Scrabble in which you end up with too many rarely used letters.

So, let us READ BETWEEN THE LINES, to go along your LIFELINE from a telephone LINE, to dropping a LINE to ask a friend to go LINE DANCING, finding your life has a SILVER LINING. But then you start ignoring the GUIDELINES, getting OUT OF LINE, and LINING your own pockets, crossing the LINE and getting into the newspaper HEADLINES. Downwards at a MAINLINE railway station to MAINLINING DRUGS. It's the END OF THE LINE, finally in hospital on life support and FLATLINING. You have met your DEADLINE!

A whole tragic novel of a life history recorded in this one word!

LOOSE WOMEN ☞ JUMPER, STRAITLACED and also RUFFIAN – because that's one place you'll find them!

Here's another phrase that was textile-linked at one time, but lost its clothing connections and, like an orphan, was taken into common usage. In the 16th century, at about the same time as the ruffians earned their sobriquet, we start to get the idea of a LOOSE WOMAN. As orphans sometimes do, the phrase has slid down in the social scale to become something more vulgar – in this case a synonym for a woman of easy virtue.

The words 'loose' and 'to loosen' come from two linked sources. From one, *lausaz*, we got words that meant undoing; through Greek they also gave us 'analyse' and 'paralyse', and through Latin, 'dissolve' and 'solution'.

The other source word, the Old Norse *slakr*, developed into Latin as *laxus* with the verb *languere*, from which we get 'to languish', and also 'lax', 'relax', 'release', 'relish', and 'relinquish.'

Perhaps not surprisingly, the even older prehistoric Germanic word *slakaz* is the probable source both of 'to slacken' and 'slacks' for loose trousers in the early 19th century, though 'slack' for the dusty residue of coal derives from a quite different Dutch source.

One of the first artificial aids to controlling a woman's shape was the *busc*, developed in 15th Century Italy, though it is likely that women of fuller figures had always indulged in a little inner garment lacing or binding. The *busc* was an elongated triangle made of unyielding material, often wood or horn, and was usually worn with a stiffened waistcoat laced at the sides in two parts, and therefore called 'a pair of bodys', from which we get our word bodice. Women who aspired to respectability, let alone fashion, would wear these extremely uncomfortable and confining garments laced up as tightly as they could bear.

An alternative and more comfortable support system allowed women to wear LOOSE JUMPS, a plural word, derived from the *jupe* – in this context a woman's underbodice. These 'jumps' were made of linen or leather and could be laced up, but they were not very stiff and could be worn comfortably slackened in a state of comparative undress.

The male author of a humorous poem of 1762 talks about these more relaxed garments. His poem contrasts the many different clothes a woman might wear according to her mood of the moment:

> "Now a shape in neat stays, now a slattern in jumps," and finishes,
> "You are hardly the same for two days together."

A folk rhyme is even more explicit in a jocular way,

> "I married my wife by the light of the moon,
> She never now gets up till noon,
> And when she gets up she is slovenly laced,
> She takes up a poker to roll out the paste."

I'd say it was lucky for him she didn't take up a poker to flatten his head!

A vulgar 18th century toast, showing masculine fixations on something far more interesting than clothing was, "Here's to both ends of the busk". Nothing much really changes. Then as now, some men believe that a woman's breasts are there for them to ogle.

So, how did women of yesteryear manage if they were on their own? A widow, a single mother, a victim of war, ill-luck or sickness faced tough challenges. If you had been one of them, you wouldn't have anyone – or at least no adult companion – who could lace up your stays to the required tightness. Perhaps you wouldn't have been able to afford such fripperies as busks and bodices anyway, and it would certainly be easier to do all your household tasks unconfined.

For such women, the absence of corseting signalled a step down in the world – literally "letting themselves go" – and they were thereby seen as slatternly and sluttish. A lone female, especially one facing loneliness and poverty, was vulnerable to sexual exploitation by opportunistic men. Her only option might be prostitution to keep herself and her children from starvation or the workhouse. So it was a slippery slope to LOOSE WOMAN meaning **of slack morals.** "She's no better than she should be!" the neighbours would exclaim, **"She's nothing but a common tart."**

Let's be charitable and admit that what separates the RUFFIANS and LOOSE WOMEN from ourselves may be no more than the relative security of our modern day social security safety net.

We still use the terms STRAIT-LACED and CORSETED for the opposite of a relaxed 21st century woman, who dresses in comfortable style. Originally, the wearing of very tightly-laced corsets would be seen as admirable and respectable – literally making you **an upright person**. Nowadays, the term STRAITLACED carries overtones of **priggishness or an excessively rigid and self-righteous morality.**

MANTLE

A MANTLE is another form of cloak or loose upper garment. The origin of the word is not certain, though probably from the Roman *mantellus*, which came to mean a 'short cloak' after starting out as a towel, napkin or tablecloth in Latin.

MANTEL may also have come to us through the Celts via the Old Norse *motull*, also meaning a cloak, and has been used in English in the metaphorical sense of **something which enshrouds or covers** (think of a gas mantel, for example) since the 13th century.

It is rather more poetic to speak of cheeks MANTLED with **a blush** or the land MANTLED with **flowers** than clothed with them. MANTLING also described **the scrollwork** behind an illustration of a heraldic shield, because it represented the cloth the knight wore under his helmet to protect his neck.

The MANTLE is also a term used in Geology as the technically correct name for **the layer of the earth between the crust and the core**, and in Zoology for **a particular outgrowth in the shell of a mollusc**. Unless you move in these somewhat rarefied academic circles you aren't very likely to find yourself needing the term.

In Spain, a traditional lacy veil, called *a mantilla* is arranged over a high comb at the back of a woman's hair so that it falls gracefully over her shoulders. The Spanish also speak of a *mantilla* as a diminished form of cape. Both come from the same root word, but it is not certain whether THE MANTLE came to us from Europe or from a Celtic linking source, and of course could have come from both. There is also a probable link with the Italian town of Mantua, famous for its silks. That town gave its name to the *mantua*, a 17th and 18th century loose gown that fell from the shoulders to the floor, worn informally at home. Later, it became an overdress with lots of tucks and a skirt short enough to allow a fancy, contrasting petticoat to be admired. The spelling of the word settled down from an earlier form of the word, to become 'mantle' in English.

A very old use of the word MANTLE can be found in the Old Testament of the Bible, where the MANTLE of Elijah the prophet symbolically and actually passed to Elisha as his successor [II Kings, chap. 13]. In this story, Elijah was taken up to heaven in a fiery chariot, but his mantle fell onto Elisha, who was later able to perform miracles with it, such as parting the waters. We use the expression THE MANTLE OF ELIJAH HAS FALLEN ON HIM figuratively to suggest someone as **a natural successor** to a great man.

Because of MANTLE'S connection with linen draped over a table or shelf, we now have a slightly different spelling for the MANTELPIECE or MANTELSHELF **over a fireplace**. Although it does not necessarily have any drapes or runners on it today, this object's name continues to remind us of its textile-related origins.

MASK ☞ DISGUISE

We use a group of figures of speech about the idea of a MASK, once a covering of rich fabric to conceal the face, and with eyeholes to peer from; later, by extension, it meant **anything that acts as a disguise**. You can MASK YOUR FEELINGS, for example, and PUT ON, ASSUME, PULL OFF or DROP A MASK to describe various ways of **concealing or revealing your expression and intentions**.

There is a similar word, also used exactly as for a MASK, and it is a MASQUE, which originally just meant the same thing. However it took a slightly different route and came to mean a **grotesque antique stone head**. From that – and rather mysteriously – it gradually changed to mean a **masked ball** or MASQUERADE; a party where you went MASKED and incognito, getting up to all sorts of naughty behaviour no doubt. To complicate things further, the word MASQUE could also mean **a particular fabric used for party clothes**. Truly a Harlequin of a word!

MITTENS

If someone offers to SHAKE YOUR MITT, don't worry. It just means they want **to shake hands with you**.

A Mitten is, basically, just a glove which covers your wrist and palm and not the fingers, or only the joints nearest the palm. When we speak of mittens we are saying literally, 'half a glove'. A Latin adjective *medietiana* meant 'cut off in the middle', and then became the noun *medietas* or 'half', from which we get moiety, medium and middle. The abbreviation 'mitt' dates from the 18th century.

It's impossible to think of anything in the least sinister about mittens; they have an irresistibly cosy sound to them. All the same, a children's rhyme tells of 'Three little Kittens who lost their mittens', and how that results in not having any pie for tea and weeping piteously. Oh, the existential angst in that simple story – and the wild elation when they find their mittens again! Who says Nursery Rhymes aren't profound?

Question: What happened to the cat who swallowed a ball of wool?

Answer: She gave birth to a litter of mittens.

Sometimes, fingerless gloves are referred to as MISERS' MITTS, perhaps because they use **less wool** than full size gloves, or maybe **make it easier to hold onto your coins** when scrabbling in a purse for pennies.

And after all this, we mustn't leave out the all-American BASEBALL MITT. The catcher stands behind Home base wearing a specially-padded mitt with all four fingers in one space and a separate section for his thumb. Players at the other bases wear a slightly different sort of mitt. And, believe me, they need protection for their hands, since a pitcher can throw a baseball at over 100 miles per hour.

MOP

Saying someone has a MOP on their heads means they have an **unruly and unkempt mass of hair**. The origins of 'mop' are obscure, but the word means, literally, a bundle of rags or coarse yarn fastened to a stick and used to clean floors or the deck of a ship. So, young man, if you innocently describe your etymologically literate girlfriend's hair as a MOP, don't be surprised if she **wipes the floor with you**, as an earlier form of this saying was, TO MOP THE FLOOR WITH SOMEONE.

MRS. MOP is a traditional slang name for a charlady or cleaner, a staple character in British fiction, typified by the lugubrious lady from the wartime radio show ITMA ['It's That Man Again'] She caused general hilarity by appearing at inappropriate moments with her double-entendre catchphrase, "Can I do you now, sir?"

"THAT'S THE WAY THE MOP FLOPS!" is a 1980's Showbiz variant on the 1950's American expression, "That's the way the cookie crumbles." It's always delivered with a shrug, as if to say, **it's just the way things are**.

MOTLEY

The term MOTLEY originated in the 14th century, and was in common use at least until the 18th, describing a woollen fabric of mixed colours, sometimes but not always chequered. This patchy "pied pattern" (yes, as in the Pied Piper or a piebald horse) was used for the traditional costume of the Fool or Court Jester, putting him outside the social hierarchy. Indeed, in the time of Queen Elizabeth I of England, the Jester was not even subject to the Sumptuary Laws, which laid down very strictly what people of different status could or couldn't wear. Nowadays, we might say, with a phrase borrowed from the American Civil Rights movement, that the Fool had the freedom "to speak truth to power".

MOTLEY became a figure of speech for **a varied mixture** of almost anything. We speak of someone playing the fool as PUTTING ON THE MOTLEY, and we use MOTLEY to mean **varying in character** as well as **parti-coloured** or

chequered. In the same way, MOTTLED and TO MOTTLE for **blotched and spotted things** comes from MOTLEY by a linguistic development known as 'back formation'.

A MOTLEY CREW is **an untidy and disreputable bunch of people**, and refers to the dubious behaviour and reputation of travelling actors and troubadours. There was even a 'glam metal' **pop group** called **Mötley Crüe** formed in 1980's America, and only disbanded in 2015.

Occasionally one hears a phrase used to encourage others in adversity – ON WITH THE MOTLEY. It means **the show has to go on, life must go on**. It was a theatrical term alluding to the Clown's cry, "vesti la giubba" ["put on the jacket!"] in Leoncavallo's opera of 1892, 'I Pagliacci.' This might be said jokingly nowadays by anyone who **has to proceed with something in spite of difficulties**.

MUFF

A MUFF is a hand warmer shaped like a short cylinder, within which the hands can be thrust. They were most fashionable in the 19th century, and are not really used now, although occasionally, fashion designers try to bring them back. In chilly climates throughout history, hands have been protected and insulated by extensions to the sleeves, separate bag-shaped warmers or various types of gloves and mittens. All sorts of materials have been used for this purpose, but fur must always have been the most popular in northern lands.

The use of muffs mainly [but not entirely] by ladies, and fur's similarity to pubic hair may have given rise to the slang term MUFF for **a woman's genitalia**, or for the **woman** herself. This came into usage by at least the late 18th century, though as is often the case it has added to in current urban slang by such terms as MUFF DIVER or MUFF MUNCHER for a partner who indulges in **cunnilingus**.

The word comes originally, in all probability, from medieval Latin, *muffula,* via Dutch, *moffel* or *moff*. The Old French *moufle* suggests a link with phrases like TO MUFFLE a sound, meaning **to mute it**. You can do this by using something soft or furry as an auditory baffle that covers either the object (e.g. a bell, oars, and the heads of drumsticks) or for the ears of the listener.

In Europe and India, **a silencer is attached to a car's exhaust system to dampen the noise** it creates; in the USA, this item is known as a MUFFLER. Interestingly, A MUFFLER also refers to a stout winter scarf wound round the neck. The same word referred to a scarf covering the head and part of the face for **concealment** or **modesty** as early as the 15th century.

A man who is a bit of a duffer at sport may MUFF A CATCH by dropping the ball. This led to the disparaging epithet of A MUFF for **a foolish or clumsy person**. This hints at the idea of being unable to catch something because you have your hands tucked snugly inside a muff.

MUFTI

A British soldier in or after the two World Wars might say he was wearing MUFTI when he was out of uniform. The expression was used, perhaps facetiously, for anyone who was **entitled to a uniform but wearing casual dress**. First recorded in English in 1816, it came from the Arabic *Mufti,* literally 'judge', but usually referring to a professional jurist and Islamic scholar who interprets Quranic law. The Grand Mufti is still the highest religious law official of the Sunni sect in Muslim countries. In 16th century Turkey, the official heads of the state religion (before Kemal Atatürk's reforms) were also called *muphtie*. They wore flowing robes, slippers, and distinctive headgear. By the early 19th century, off-duty or retired army officers affected MUFTI by lounging around in a costume of a silk smoking jacket, slippers and a tasselled fez – a style later used theatrically **to denote exotic eastern foreigners**. Eventually, any military or official person who chose to dress in '**civvies**'; **the casual style of a civilian** rather than a soldier, could say he or she was IN MUFTI.

NAKEDNESS ☞ BIRTHDAY SUIT

Let's take a quick peek at nakedness. Our word **naked** goes back, like much of the human race itself, to an Indo-European word *nogw*, for nakedness, from which the Romans got *nudus*, hence 'nude'. Various European cultures transformed it in slightly different, but still recognisable ways.

Nakedness is often seen as the earliest state of the world and there is an Australian aboriginal story that tells how Father Sky and Mother Earth were originally naked and so close together that their children had no space to breathe and had to wriggle and push to separate their parents. Father Sky looked down and was overwhelmed with how beautiful Mother Earth was but all brown and naked, so he created ornaments out of dust and sunlight, scattered them over her, and they became THE CLOTHING of greenery and trees that now cover the earth.

Going NAKED is also called '**wearing your Birthday Suit**' because, of course, we are all born 'without a stitch of clothing', to use another phrase for nudity. Sometimes the dramatic power or shock value of nakedness is enhanced with modifiers, e.g. STARK NAKED or BUCK NAKED.

A modern variation is GOING COMMANDO, meaning **to go without underwear**, usually in the crotch area. It was not known before the second half of the 20th century, and is of uncertain origin. It may come originally from female prostitutes, or from the US army in Vietnam, where underpants caused problems in a hot climate, and were therefore discarded.

The phrase The NAKED TRUTH comes from a fable in which Truth and Falsehood went swimming. When Truth returned she found Falsehood had stolen her clothes. She refused to wear THE GARMENTS OF FALSEHOOD and had to go naked thereafter.

Nakedness also meant vulnerability, and was a way of shaming and humiliating an enemy. An evil Maharajah told his servants to strip naked a woman who had refused his advances. In her shame, as the men started to unwind her sari, she prayed to Lord Krishna, and he heard her prayer. The sari never ceased to unroll and the court was filled with miles of billowing silk, until even the Maharajah was frightened and let her go unharmed. I wonder what happened to all that silk? I hope she was allowed to keep it.

We may speak of THE EMPEROR'S NEW CLOTHES. This alludes to a story from Hans Christian Andersen, and the phrase has passed into the language to suggest **a person who doesn't realise what a fool he is making of himself**, or is **generally unaware of how he appears to others**. A vain and foolish Emperor is deceived by two clever con men. They pretend to make him a wondrous, magical suit that is invisible to anyone stupid. Of course, neither the Emperor himself nor his courtiers want to admit to being stupid, so they all pretend to admire this non-existent clothing. When the Emperor parades through the town in the nude to show off his wonderful new outfit none of his subjects dares to say anything, until a child bursts out laughing and cries, "He's got nothing on!" and then everyone finally admits the truth.

It was originally a folktale which Hans Christian Andersen rewrote to play with ideas about nakedness and clothing, showing how both the Emperor and the crowds of sycophants around him have the ability to fool themselves to truths which may be patently obvious to the innocent and naive.

We also play with the story of Adam and Eve, cast out from the Garden of Eden naked except for some greenery. Here is a Vicar's announcement:

> "There will be an Adam and Eve social next Thursday. Starts at 7 – leaves off at 11."

NIGHTCAP ☞ CAP

A NIGHTCAP was a very necessary article of clothing in cold countries with draughty bedrooms, as it was a padded cotton or woollen head covering, usually fastened with strings tied under the chin so that it was a snug fit.

Such headwear has been known by this term from at least the 14th century. Then, around the 17th century, NIGHTCAP became a term for **a good night drink**, usually alcoholic, to be quaffed for warmth before bedtime. Indeed, a second one was sometimes referred to facetiously as '**a string to tie it with**.'

In modern slang, it has been used between young people to ask if their date is going to "come in for a nightcap?" This is a deliberate use of an old-fashioned word as a euphemism to suggest **a sexual encounter**.

NIGHTDRESS

The NIGHTDRESS, also known as a NIGHTGOWN, is as straightforward as it sounds: a loose dress worn at night. It was usually for women and children, and its name was shortened to A NIGHTIE or NIGHTY. Men tended to wear a nightshirt. To describe someone's day clothes as "Like a Nun's nightie", would be an insult, as this Irish saying means their garments are **crumpled and ill-fitting**.

The nightdress has given rise to the phrase, NIGHT-NIGHT, or NIGHTIE NIGHT, when **saying goodnight to a child**, as it is associated with getting a child washed, warm and cosy in their nightwear, then giving them a loving hug and kiss at bedtime. The saying has been in use since the late 19th century, and like so many things in our culture, particularly to do with children, its usage has never really been discussed; it was simply absorbed by the English speaking world. NIGHTIE NIGHT is an affectionate term and can be used between friends, though usually with a slightly jokey sense of its childish connotations. In recent times, it sometimes crops up in a more sinister way in gangster films, at the moment when **someone is knocked unconscious**.

In bygone days, there were children's rhymes about this universal garment:

> Wee Willie Winkie runs through the town
> Upstairs and downstairs in his nightgown.
> Rapping at the windows, crying through the lock,
> "Are the children all in bed, it's past 8 o'clock?"

No good asking anyone that question today. They'll only wonder why Willy Winkie is running about at all hours in a state of undress, if not sniggering at his name. They might also wonder why some annoying person is singing a rap through their windows, and be ready to ring the Neighbourhood Watch to cart them away!

This is an example of one of the few nursery rhymes that were actually written for children and did not start as political or satirical verse. It was in *Whistle Binkie, a collection of Songs,* by William Miller, published in the mid-19th century.

PALLIATIVE ☞ CLOAK and VEIL

Celtic legends tell of the legendary Old Woman, their archetypal Wise Woman. She is called the Cailleach, and her name means 'THE VEILED ONE'. The word has transmuted a great deal, but nevertheless, the Latin word *pallium* for cloak is thought is the origin of this and a number of related terms. The same root gave us PALLIATIVE, as in PALLIATIVE MEDICINE, which "cloaks" or **mutes pain and other symptoms of serious illnesses**.

A PALL is also a little **square of linen** used to cover the Chalice in some churches at the Communion service. The word has also been used since the 15th century for **the covering over a coffin**, often richly embroidered, because in early Christianity in Rome the Christians wore a purple robe rather than the conventional toga. *Pallium* was a cloak or covering and the *palla* was the long upper garment worn by Roman women.

This link with **gloominess**, from the association with the dead, may have led to the use of the word PALL in the 18th century to mean **to fill someone with horror or dread**, or even just getting **bored**, "THAT ACTIVITY HAS BEGUN TO PALL ON ME". We can be stronger and describe something as APPALLING. "WHAT YOU SAID APPALLED ME", or "THEY BEHAVED APPALLINGLY". is **a very strong condemnation**.

PALTRY ☞ ROBE and RUBBISH

If you were to give me a tip for some service I had done for you, I might look at it and say sniffily, THAT'S A PALTRY SUM! [No, no; I wouldn't. Just try me!]

We use the term perhaps most about a sum of money today, but it has always been used to express **a sense of disdain at something insignificant, meagre, rubbishy, useless, measly, miserable, puny, piddling, derisory, petty, pitiful, miserly, trivial, inconsequential, trifling, beggarly, wretched, slight, ineffectual and unwanted**. It's not as if we are short of rude synonyms to express disdain!

This one, however, really is derived from textiles, coming to us from various sources. *Palter* in the 16th century was Old German for 'a rag', and *paltrig* similarly meant 'rubbish'. There are Scandinavian words which suggest the same idea, such as *spalten*, to split off, as a piece torn or cut from cloth, a useless cut-off, a rag.

PANTS ☞ BREECHES and TROUSERS

There are remnants of original underwear dating from over 7,000 years ago, and these appear to have been flat leather strips that could be wrapped around the groin, as a loincloth still is in some parts of the world. They were, presumably, worn by men, since dainty flower-printed knickers for women weren't to be invented for several millennia. The Romans called such a garment a *subligaculum* which could be either a form of drawers or a loincloth twisted around.

PANTS is short for pantaloons, a word coined from old man Pantalone in the Italian 18th Century Commedia dell'Arte, who symbolises **money, acquisitiveness** and **greed**. His name derives from *San Pantalone*, a popular saint in Venice. Oddly enough, he didn't wear loose pantaloons as you might expect, but tights. In Victorian times, pantalettes, bloomers and drawers were all forms of underwear for ladies; pantalettes were long and often had pretty lace around the ankles that was meant to be seen below the hem of a skirt.

However, as we've already discovered, the same words can have very different meanings when they migrate across the Atlantic. In America, PANTS – also called 'slacks' – refers to TROUSERS, while the same word still means knickers or underpants in the UK. Trying TO SCARE THE PANTS OFF HIM, or TO BORE THE PANTS OFF HIM, or even TO BEAT THE PANTS OFF HIM, are all fairly obvious, and use the US sense of the word; the last usually means **to beat someone soundly in a game**.

Despite their Yankee origins, all these sayings have been adopted enthusiastically by the British. If some notable person PUTS ON HIS PANTS ONE LEG AT A TIME, an American would understand that to mean that **he's an "ordinary Joe", like anybody else**.

In modern British slang, "THAT'S PANTS!" means the situation or object under discussion is **really rubbish** or simply **not fair**; for example, "Tuition fees are going up again this term. THAT'S JUST PANTS!"

It is not the garment itself, but the area it normally covers, that provides the force for taunting someone, "PANTS TO YOU!" – a common playground jibe. You can add another word if you want, such as shouting SMARTY-PANTS at

a clever person who seems to be showing off a bit, and SCAREDY-PANTS or 'FRAIDY-PANTS to **someone who won't accept a dare**.

You might chant, "LIAR, LIAR, PANTS ON FIRE!" which adds rhyming for emphasis and extra punch to deride **someone whose words you don't want to believe**.

Underwear has always produced giggles and sneers in the playground, with the possible humiliation, during a particularly active game, of showing your bum unintentionally. Combinations, a rather old-fashioned term now, were a singlet or vest for a man combined with long underpants. Here's a derogatory First World War song children used to chant:

> "Kaiser Bill went up a hill
> To conquer all the nations.
> Kaiser Bill fell down the hill
> And split his combinations."

Here's a lively counting-out or rope-skipping rhyme, recorded in a modern Australian playground:

> "Cinderella, dressed in yella
> Went upstairs to meet a fella.
> On the way her pink pants busted,
> How many people were disgusted? 1, 2, 3, 4, 5, 6."

Grown-ups are always saying gloomy things like, "Children today aren't as free as we were!" Were they ever really free? An extended innocent childhood has usually been the privilege of a very thin slice of the demographic cake. Throughout history, in most of the world, children have had to grow up fast and work hard for their living. Many still do.

PETTICOAT ☞ BREECHES and COAT

The simple word 'coat' has rather a complicated, not to say sexually-ambiguous past. Not suprisingly, it is also very ancient, and has evolved both as a useful practical garment and as a versatile metaphor.

To English folk of the 13th century, it was a 'cote.' This links to PETTICOAT – a unisex word meaning petty or 'little coat'; in other words, a short coat. Petticoats, in the plural, went their own sweet way, as you would expect of something so feminine, to become a woman's undergarment in the 15th century. At that time, the loose clothing of children was also called a petticoat, and a boy would wear only this garb – just like his sister – until he was dressed in men's breeches, usually at about the age of six years. Since about the end of the 16th century, the word PETTICOAT has been regarded as such a typically female garment, that it bestowed its name to the whole sex.

The petticoat also appears in folklore, with associations similar to any intimate garment, which can **stand for the whole person** or **carry the essence of that individual for magical purposes** – in the same way as hair or nail clippings might be used. A common European folk belief was that a pregnant woman, seeing a hare, or stepping over its lair, could cause the facial deformity of a cleft lip known as "harelip". However, she could easily break this curse by tearing or cutting a strip off her petticoat or dress. Not all girls were willing to do this, as clothing was cherished for years or decades in those days, unlike our "wear-for-one-season-only" modern culture. Research into this countryside belief as recently as 1981 in the UK showed that six older women in a survey of 200 still believed in this superstition!

Women who take a leading role either at home or in public life have often been regarded as a threat to men – and sometimes to other women. Their activities have been described disparagingly as PETTICOAT GOVERNMENT. The original use of the phrase seems to be obscure, but it was used by Baroness Orczy, author of *The Scarlet Pimpernel*, as the title of a novel about the influence of Madame de Pompadour in the 18th century French Court. The book came out in 1910, and was reprinted in the USA as PETTICOAT RULE.

A more modern use of a similar phrase, PETTICOAT or PINAFORE DISCIPLINE describes **an erotic practice** in which men are 'forced' to wear female clothing and humiliated – although, of course, they put themselves voluntarily into this position.

There is a public thoroughfare in London's East End that runs between Middlesex and Goulston Street, known since the mid-18th century as PETTICOAT LANE. It is the location of a famous **Sunday market for clothes**. This area of London, known as Spitalfields, was the former site of the medieval St Mary's hospital and priory, a place where travellers, the poor and the ill received "hospitality" and care. Later, the neighbourhood attracted Huguenot silk weavers fleeing persecution in 17th century France, and, in time, both Irish immigrants and Eastern European Jews working in the clothing industry settled here. Today's Petticoat Lane market mingles tourists and locals alike, and is still the place to go for fashion bargains.

Here is a Skipping Rhyme from Australia:

> All in together
> This cold weather.
> I saw a nannygoat

> Putting on a petticoat.
> Shoot! Bang! Fire!
> All run out.

And a Riddle Rhyme from the UK, at least three centuries old:

> Little Nanny Etticoat
> With a white petticoat
> And a red nose.
> She has no feet or hands,
> The longer she stands
> The shorter she grows.

Question: What is she ? Answer: A lighted Candle.

PICCADILLY ☞ HABERDASHERY

About five centuries ago, there lived a London tailor called Robert Baker. He made a lot of money and decided to build a grand, prestigious house for his family. He really wanted to locate it within the square mile of the City of London and near the River Thames. However, at that time, the City had a wall around it. The good Burghers, Aristocrats and Members of Craft Guilds who lived inside its well-protected area had decided their patch was already too crowded, so they turned away the upstart Robert Baker. Thwarted, he bought a plot of pretty, green land to the west of the City in 1612. There he built a modest mansion and called it Baker Hall. Then he bought up other packets of land nearby and became a land speculator and (aspiring) Gentleman. He often got involved in litigation, as his building activities were of dubious legality. They also gave rise to lots of infighting among his heirs over the next 60-odd years following Baker's fairly early death in 1623.

Robert's wealth was founded on the sale of what we now call HABERDASHERY [qv]. In particular, he was a major importer and seller of small, stand-up lace collars, often made in Italy of the famous reticella lace. They were called PICKADILS, probably from the term *pic* or *pike;* a Turkish measure of length used for cloth first recorded in 1579, according to the Shorter Oxford English Dictionary.

Baker's collars were usually narrow bands or semicircles of lace stiffened with simple pasteboard strips. Indeed, these strips also became known as PICCADILS. They were the sort of irritating little accessory, like cufflinks and shirt studs in a later age, that often fell down behind the wainscoting and had to be replaced. Very probably, that's why he sold such a lot of them!

One way and another, our Robert was well known to have made his money buying and selling these very humble items, and all his airs and graces wouldn't disguise that fact. If a newcomer to the vicinity asked, "And where might I find the famous Baker Hall, domain of the Gentleman, Robert Baker?" the local guttersnipes and ragamuffins would reply, "You'd be meaning Piccadil Mansion, wouldn't you Guv?" [or the 17th century equivalent] and then fall about sniggering, with their grubby fingers stuffed in their mouths. So poor Robert had to suffer the ignominy of a nickname for his house that exposed his 'trade' origins to the end of his days.

Well, that's the story and it's far too good not to be true! The word PICKADIL subsequently came to be used for a stuffed, rounded hem on a skirt, and could also refer to a house on the farthest outskirts of an expanding city such as London. By the 1630's, **the whole area of Soho around Baker's mansion was known as Pickadille.** Today, long after Robert departed this life to sell lacy collars to the angels, the Street of PICCADILLY and its famous London landmark of PICCADILLY CIRCUS proclaim their textile origin.

By the by, Robert Baker's house may have stood on the site of the current Lyric Theatre, which featured in Bram Stoker's famous novel, 'Dracula' written in 1897. This was the manse taken over by Count Dracula for his nefarious (or do I mean 'nosferatu'?) plans, who chose it because it was "in the heart of fashionable London."

PIG IN A POKE ☞ POCKET, PURSE and SACK

A POKE was a drawstring bag, or a cone-shaped twist of paper used to hold small goods, such as loose sweets, and it could also be a SACK [qv]. The word is not related to 'poke', as in to thrust with a knife, stick or finger, but probably from very early Germanic *puk*, meaning a bag or pouch. In the 13th century, Old English had it as *pohha* or *pocca* and French as *poque* or *poche*. Similar words occurred elsewhere in Europe, all roughly denoting a bag, purse, pouch, or pocket.

There is a story which is more of fable than a true occurrence, though it is often repeated because of its useful lesson. At country fairs, you might be offered a pig sight unseen, particularly if it was on the black market ("It fell off ye back of a tumbril, yer honour."). When you got it home, if you had been fool enough not to look into the sack before handing over your money, you might find you'd BOUGHT A PIG IN A POKE and **been palmed off with a cat**, hence also (because that angry cat wouldn't stay put) TO LET THE CAT OUT OF THE BAG.

Frankly, I think this is a bit silly. Cats and pigs, even very small farrows, have a very different outline and weight, even inside a sack. They make different sounds entirely and cats use their claws to try to break free, while pigs only kick and wriggle. I think the phrase may have come from the practice of drowning unwanted cats and kittens in a sack. Simple as that. If you opened the sack before they were properly drowned, they'd scarper. A wise old lady like me, if I had met you selling dubious goods at a 14th century fair, would have uttered the following sage pronouncement, "When me proferreth ye pigge, opon thy pogh."

PILLOW ☞ CURTAIN

The word PILLOW seems to have been coined in the 14th century from Old English *pyle,* for a soft, fabric-covered cushion on which to rest your head, particularly in sleep. It is therefore very similar in use to a BOLSTER. This larger, longer version came originally from a Germanic series of words around the ideas of stuffing and of swelling: *bolg, bulg, bolstraz, polster* – which eventually gave us BELL, BELLY, BELLOW, BILLOW, BULGE and BOLD.

A PILLOW DOG nowadays **is the sort of little, fluffy canine friend that sits on a lap or pillow** and is so beloved by certain types of film star as their **accessory-cum-companion** [see HANDBAG DOG]. In earlier times of lesser hygiene, these dogs had a very useful purpose. They would sleep beside the head of their master or mistress all night in snoring contentment. Because a dog's resting temperature is higher than that of a human, in the night, all the little bugs, fleas, nits and other itchy nasties would migrate into the dog's fur (in theory), giving their owners a welcome respite. I'm scratching as I think of it. "Here, Fido, come here!"

PILLOW TALK is **the quiet conversation that goes on between partners** at the end of the day, often in the intimate darkness of the bedroom. It is one of the rare and lovely pleasures of a happy relationship and is totally the opposite of A CURTAIN LECTURE.

PILLOW TALK is also used to refer **to disclosures of privy information**, such as by government officials or military personnel who dally with playmates later disclosed as journalists or undercover agents (literally under the covers!) Think of 1960's model Christine Keeler and John Profumo, Secretary of State for War in the 'Cold War' era. Another of Keeler's "boyfriends" was a Soviet spy. Profumo's pillow talk not only caused a huge scandal, but also forced both his resignation and that of Prime Minister Harold Macmillan a few months later. All of which goes to show that, if you indulge in indiscreet chatter to dangerous companions, even the softest PILLOW will not **cushion the blow**.

A PILLOW BOOK refers to a Japanese **'love manual'**, a kind of diary to be kept on the pillow with sex tips, poetry and even woodcut illustrations, originating in 10th century Japan.

POACHING ☞ POCKET and PURSE

From the French *poche* we got 'pouch', and the French verb *pocher*, meaning 'to put in a bag' gave us the idea of POACHING something, whether **stealing animals or ideas** as well as POACHING EGGS by swirling boiling water and a drop of vinegar with a spoon and then slipping a raw egg into this mini-maelstrom so that it wraps itself up inside a little bubbling pocket of egg white and cooks to perfection.

It seems likely that the POACHING which refers to **illicitly killing wild animals and game birds** came about because you would have to hide the proceeds in your pockets or a bag.

Small creatures such as rabbits and pheasants were forbidden to the poor because the aristocracy had enclosed so much common land in the 18th century. Ordinary folk naturally felt aggrieved, retaliating by taking what they felt they were entitled to, in order to feed their starving families. We still use the term POACHING for **any depredation of wildlife**, such as the terrible killing of elephants for their tusks.

POCKET ☞ PIG IN A POKE and PURSE

This word derives from the same roots as POKE and PURSE above. It then adapted itself by becoming integral to the clothes we wear, rather than a separate 'bag' of some sort that can be carried. But the distinction wasn't always that clear. In the 17th and 18th centuries, pockets were pear-shaped bags slung round the waist, often under clothing and accessed by slits in the outer garment. It is this sort of pocket that Lucy Locket famously lost:

> Lucy Locket lost her pocket,
> Kitty Fisher found it,
> Not a penny was there in it,
> But the ribbon round it.

Lucy Locket was a general name to suggest a young woman, but Kitty Fisher was a historical person, Georgian beauty and daughter of a milliner, who made her way up in the world via various bedchambers, eventually marrying John Norris, Gentleman. Her portrait hangs in the 'Room of Beauties' in Petworth House, Sussex. It is said that, during her career as a courtesan, Casanova offered her £10, but she refused him, as she never took less than

£15. She never forgot her roots and was generous to the poor. She sounds like a spirited lass, and it is sad that she died quite early, of smallpox.

To be OUT OF POCKET is **to fail to get back money y**ou have spent that you expected to have returned, and TO POCKET SOMETHING is slang for **stealing or embezzling**, even if the booty is not literally placed inside a pocket. You may have TO POCKET YOUR PRIDE or even POCKET AN INSULT, meaning **to accept things without apparent rancour**, although you're probably seething inside.

TO HAVE SOMEONE IN YOUR POCKET is **to have power over them**, with a hint of dishonesty. TO LIVE IN HIS, OR HER, POCKET is **to be tiresomely dependent on the other person**, who is around all the time. LINING YOUR POCKETS again implies a measure of dishonesty, or at least opportunism, as it suggests **you are tucking away a lot of money into said pocket**s – or more likely today, into your Swiss bank balance.

POCKET MONEY is **a small sum** such as children might be given for sweets, or a derisory offer of wages. In fact, anything small may be described figuratively as if it had been **designed to slip into a pocket**. It could be anything from A POCKET EDITION of a book to A POCKET BATTLESHIP. The latter refers to the way the German shipbuilding industry tried to get round the restrictions on the size of ships they were allowed by international agreements to build after the first World War. They were smaller, but certainly not diminutive, and did quite a lot of damage. I'm not sure I like the idea of slipping a whole battleship into my pocket anyway; it might strain the seams and would certainly spoil the line of my outfit.

The nicest metaphor hints at something many women would like to be: A POCKET VENUS; it means you are **petite and exquisite**. Or as modern parlance has it, **small but perfectly formed**.

To have DEEP POCKETS is **to be rich**. But even being very wealthy wouldn't save you from this rhyme's tale of doom:

> "Ring-a-ring-o'roses,
> A pocket full of posies,
> Atish-oo! Atish-oo!,
> We all fall down."

The pocketful of posies is often said to be the sweet-smelling herbs used to avert the plague, and the rhyme to be about the Black Death, but there are those who think that it was just a game that was used to get round the Puritan ban on dancing, with a curtsy at the end.

PORTMANTEAU ☞ MANTLE

This was originally a container or case in which to carry your mantle, from combining the French for *manteau* and *porter*, 'to carry'. It was also once the title of **the court official** who carried the King's cloak, as well as the bag in which it was transported, and thus it became a general term for **a travel bag**. A suitcase fulfills the same purpose now, and may contain almost anything – except, in these days of casual wear, a suit.

The versatile word PORTMANTEAU eventually expanded (just like its original namesake) to become **a broad compendium of things**, such as a big collection of poetry in one book. A pity, really that we can't go back to earlier times; I like the idea of someone whose only job is to carry my coat around – or perhaps a book of poetry as well, on holidays.

PURSE ☞ POCKET

The idea of puckering or pursing the lips helps us to visualise something gathered up into little wrinkles or pockets, like fabric. In Middle English both words could signify a fold or pleat in fabric, and many words may be derived from this, such as pock, poke, pouch and pucker. To PUCKER UP means **to express disapproval** as if making a small mean mouth – although something very sour might make you pucker up involuntarily. However, TO PUCKER UP FOR A KISS, or TO PURSE YOUR LIPS, simply means **drawing them together like the gathers of a purse**.

A PURSER was originally **a maker of purses**, up until the 17th century, but later became **an official who managed the money**, in a variety of government or academic establishments, and on board ocean liners. On a merchant ship, he or she not only looks after the finances, but is chief administrator, with oversight of all the cooks and stewards.

The national treasury for Britain, known as THE PUBLIC PURSE is, of course, in the care of our friendly CHANCELLOR OF THE EXCHEQUER, and he keeps all our money in a BUDGET, a word derived from an old word for bag or womb.

{See ***Rigmaroles and Ragamuffins*** under *Exchequer*}

PURSE probably came originally from the Greek *bursa* for leather or skin, and in Latin, the same word meant a small pouch. In time, the word made its way to places of higher learning, and denoted **the money held for the purpose of funding a scholarship**, known as a BURSARY. **The man who looks after the finances**, or the PURSE-BEARER of such an institution, later became known as THE BURSAR, a title still used in English Public Schools and Universities. And where might this Mr Moneybags be found? Well, in Latin, *bursaria* is the name for the treasurer's or bursar's room – from which he DISBURSES or literally 'unbags' and pays out his largesse; in English, we again use the word BURSARY for this office.

TO HOLD THE PURSE STRINGS is **to control the money**, and an official in charge may TIGHTEN or LOOSEN THE PURSE STRINGS, as he feels appropriate, **to hold on to, or release**, the cash. Nowadays, everything is done at a distance or by electronic transfer.

IF I HANG ON YOUR PURSE I'm **living off you**, or at least **expecting you to support me**. [Oh please, please do!] The best I can probably hope for is that we will SHARE THE COMMON PURSE by **pooling our expenses**.

You might not be much use at supporting me if you have A LIGHT PURSE, as it means you are **poor**, whereas A HEAVY PURSE or A LONG PURSE means **you are well off**. The latter is an apt description of some of the early purses, which were **sausage shaped**, with a slit to get the money in and out. Such purses had two rings that could be pushed along to close the opening or separate your copper coins from the gold and silver; this was also known as a miser's purse. Other purses were little bags of silk within a stronger leather purse, and were plump and round, rather like some sort of lovely bun or pudding.

In boxing matches and horse racing circles, one still talks about A RICH PURSE if **the prize money is good** and the owner of the winning nag collects THE PURSE, even though it's probably a cheque now, or a bank transfer, and not a real purse at all.

Shakespeare's *Othello* has Iago exaggerate the value of his good name by contrasting it with money in a phrase that has become almost a proverb, "Who steals my purse steals trash."

> "Who steals my purse steals trash; 'tis something, nothing;
> 'Twas mine, 'tis his, and has been slave to thousands;
> But he that filches from me my good name
> Robs me of that which not enriches him,
> And makes me poor indeed."

The irony, of course, is that Iago, is far from being an honourable man.

"YOU CAN'T MAKE A SILK PURSE OUT OF A SOW'S EAR!" This proverb means that **you cannot make something good out of inferior materials**, either literally or figuratively. It comes from the 16th century, and the swine was chosen for this metaphor because you can apparently make something useful out of every bit of a pig except the 'Oink!'

The saying certainly holds good for textile crafts. It is therefore rather pleasing that the firm of Arthur D Little, specialising in artificial fibres decided to take on the challenge in 1921. In fact, they made two. Their chemistry laboratory extracted the protein from the pig's ear and were able to treat it to make a very fine thread. This was then woven, embroidered and sewn to make a most respectable purse, though the silk was remarked to be not of the finest quality [sighs of relief from all those silkworms, fearing redundancy].

Although some commentators remarked, rather sniffily, that the expense and difficulty of doing so were sufficient to make it impractical for commercial purposes, the aim of the project had been to loosen up expectations and use the sense of playful creativity which is behind all real scientific endeavour. The result can still be seen in the archives of the Massachusetts Institute of Technology.

In our age of refined technology, such a story is not inherently unlikely – but you may be tempted to exclaim, "IN A PIG'S EAR!" as **an exclamation of disbelief**.

PYJAMAS

Our nightwear known as pyjamas or pajamas comes from the Hindi *paejama,* itself derived from the Persian *pai* meaning a leg, and *jamah,* or garment. They are long, loose trousers still worn in India, North Africa and the Middle East and were adopted by European colonists, who added the jacket for propriety and brought the fashion home to their chillier home countries. Initially, 17th century gentlemen wore them for relaxing, e.g. when lounging around smoking that new-fangled stuff, tobacco. But probably because of all the draughts in British houses, pyjamas found their way into the bedroom as nightwear. 'Onesies' had to wait a bit longer to be adopted; but my goodness, don't they keep one's toes nice and toasty!

The idea of describing **something that was superlative** as THE CAT'S PAJAMAS was coined in the Roaring Twenties. Other similarly enthusiastic terms were the cat's meow and the cat's whiskers, suggesting the smug, self-satisfied expression on a cat's face. In fact, most of the animal kingdom seems to have been drafted into our language during that jazzy, anything-goes era, including the bee's knees and the snake's hips. Unfortunately, only the cats got to dress up in anything textile-related, so we shall have to leave the rest of the excellent (but nude) menagerie where it is.

RETAIL ☞ *Rigmaroles and Ragamuffins* for a fuller discussion

The RETAIL TRADE is that in which individual goods, or small amounts are sold to the public; the opposite of the Wholesale trade in bulk dealing. The word came from Old French *retaille* which was – literally – a piece cut off, hence originally denoting textiles of some kind. This trade in bits of cloth gave us the English TAILOR and the term 'cut' to describe the appearance and outline of an outfit. The Italians took over the idea of selling small quantities or cut-offs and the term became RETAIL for the trade in any single item or small number.

By the end of the 16th century, the subsidiary meaning of 'tell', or 'to relate' – in the sense of repeating a story – had appeared for RETAIL. It's almost as if you are 'selling' someone a juicy piece of gossip. Did I mention that I know about some really lurid scandals? Buy two, get one free!

The same idea of 'something cut off' gives us DETAIL as well, meaning **a small thing**, **a part** of a larger picture.

RIBBON

A RIBBON is now a strip of cloth, usually a few inches wide and of any length. It started as a stripe in woven material and the word is probably early Germanic, from a root similar to a band. In French it became a *ruban*, and in English a riband. {see also *Rigmaroles and Ragamuffins* under BAND for more details}

These interesting strips of fabric have many uses, some ornamental, such as coloured ribbons on clothing or hair, and others that are practical, like tying up boxes and closing bags. A typewriter ribbon – remember them in the Olden Days before computers? – was soaked in ink, and ran from reel to reel behind the striking area, where keys rose up to mark letters onto the paper.

The term RIBBON has been used metaphorically for anything that can be **made into strips**, so one might say TORN TO RIBBONS. They may also be used as a form of identification or a sign of some honour on uniforms, and to express solidarity with a group, such as the red ribbons worn to express support for AIDS sufferers, or the pink ones for breast cancer.

Since the 1990's, RIBBON has become urban slang for **the female genitalia**. It can also suggest **a pointless award** given to participants in a competition just for turning up.

An ancient Norse legend tells of a very special RIBBON, called "Gleipnir" that was the only thing strong enough to bind the mighty wolf, Fenrir. Metal chains had failed to fetter the animal, so the gods asked the dwarves, master blacksmiths, to fashion Gleipnir out of six items: the sound of a cat's footfall; the beard of a woman; the roots of a mountain; the sinews of a bear; the breath of a fish; and the spittle of a bird. Even then, Fenrir only agreed to being bound if he could hold the hand of the great god Tyr in his mouth. When he realised he had been tricked and would never get loose, he bit Tyr's hand off. Not exactly the sort of pretty story you expect to go with the lovely, colourful ribbons with which we adorn ourselves.

ROBE ☞ RUBBISH [that's a bit harsh, isn't it?]

The word ROBE, now a poetic or archaic word for a woman's dress or a description of ceremonial attire, has a fascinating and wicked past. It originated in 'rob' and meant stolen clothes, since rich fabrics were a special sort of booty. Intriguingly, a probable lost Latin word *rauba* gave us the Germanic words for both robbed and bereaved – perhaps because death robs us of our loved ones and companions. In Old French, ROBE completely lost the sense of clothing and came to mean **any odds and ends that had been stolen or looted**.

> Hark, hark, the dogs do bark,
> The beggars are coming to town,
> Some in rags and some in tags,
> And one in a velvet gown.

The velvet gown was almost certainly stolen. This rhyme probably dates from the Tudor era, when the problem of unemployment became particularly bad for all sorts of reasons. Gangs of beggars roaming the land were often desperate and violent men and women, and punishments for theft were very severe.

Women used to speak of their dresses as robes, and we might use ROBED IN BEAUTY, or in GLORY or PEACE to extol the loveliness **of women or angels**. [Aren't they the same thing, really?]

Here is a poetic little parable from the Muslim Sufi tradition, about how robes came to be used as symbols of ritual. A shipwrecked traveller in a robe found a group of other stranded people. They were naked or reduced to rags, and had poor short-term memory because of their ordeal. Somehow, he made special robes for each of them too and gave them duties and responsibilities to help them all in their new lives. Some of the robes marked the owners out for authority and distinction. Unfortunately, many of those who wore them made no effort to learn and remember their new duties and station in life, and the rituals he had devised, so the robes became empty of meaning.

The robe also gave us the word WARDROBE, which was originally a whole **room where clothes could be guarded** by a warden. These eventually shrank into freestanding pieces of furniture or lined a bedroom wall with BICs – built in cupboards. Recently, however, the emerging "must have" status symbol for Ladies Who Shop is a walk-in wardrobe with room to stash a vast array of garments and accessories.

Today, aside from the common BATHROBE many of us don fresh from the shower or at bedtime, it is mostly officials of various kinds who wear robes. Why not go up to a High Court Judge in his splendid robes and say to him, "Do you realise you are wearing stolen clothes?"

RUBBISH ☞ ROBE

As seen in the previous entry, the word *rauba* gave rise to both robbed and bereaved. Still later, it lost its connection with stolen clothing, and *rauba* developed into the Anglo-Norman *robel*, meaning **bits of broken stone**, a sort of cousin word to ROBE.

Robel came to us as 'rubble', and its plural would have been *robeus*, probably giving us RUBBISH. That's how most of us are inclined to dismiss the contents of our wardrobes when longing for something new. "I haven't a thing to wear," we cry. "It's all RUBBISH!"

RUFFIAN ☞ LOOSE WOMEN

The origins of the word RUFFIAN – and, indeed, of ruff and ruffle – are obscure. There is even disagreement as to whether or not they all come from a common source and whether this root is linked to **rough**. All that seems certain is that both the ruched and crimped collars that developed into the ever higher and more elaborate ruffs of fashion, and the word ruffian, meaning a swaggering villain with long hair and a wild appearance, arose during the middle of the 16th century. Rufflers, now an obsolete epithet, were a class of vagabonds around the same time.

The Sumptuary laws, which began in the mid-13th century and continued for 300 years, laid down the rules for all kinds and classes of person concerning colour, style, cloth and all aspects of dress they were permitted to wear. Among them, in 1562, were some strong prohibitions against the burgeoning RUFFS that had begun to exceed all decorum. The law (as ever on the side of protecting the privileges of the Top People) ordered hosiers, tailors and cutlers to modify and keep in check their ruffs, hose and length of their swords: "Neither with any shirts having double ruffs either at the collar or sleeves, which ruffs shall not be worn other than single, and the singleness was to be used in due and mean sort, as was orderly and comely used before the coming in of the outrageous double ruffs which now of late are crept in."

Even institutions of higher learning were involved. Presumably then, as now, students vied with one another in both fashionable outrageousness and rebellious attitudes. University Chancellors and Vice-Chancellors were commanded to see that their scholars "abate their ruffs and other unseemly excesses". Fat chance of that happening!

Used as a verb, **ruff** once meant to swagger, bluster, domineer and "set cockahoupe". It also meant to heckle flax – that is, to draw out the fibres to straighten and clean them – with an instrument known as a **ruffer**. Words for beating and combing flax to make thread were linked to the comb-like formation of a bird's hackles when it was ruffled, so it seems perfectly possible to me that both the terms and the vogue for showy neckwear started with game birds set to fight against each other – a popular pastime for gamblers down the ages. The birds would fluff up their hackle feathers in a frill to look bigger and more impressive and swagger up to each other aggressively. Hmm. Doesn't that remind you of certain human beings just a bit?

We speak of being **made angry or upset** as being RUFFLED, which seems likely to be from this same source, so the RUFF as the feathered collar round a bird's neck may well have preceded and given rise to **the fashion in dress**.

An outdated proverb spoke of the evil temper of "a Cook-Ruffian, able to scold the devil out of his feathers", and the Italian Commedia dell'Arte even had an old woman as a character known as Ruffiana.

A man who let his hair and beard grow long would appear as if wearing a fantastic ruff – especially if it was curly – and might also look very intimidating. The term **ruffian** was almost always used in condemnation and linked with the appearance of hair and clothing, marking the men out as dissolute **thieves** and **cutthroats**, as described in 1603, "all in their apparel as roisterers and ruffians." Oddly enough, it could also occasionally be used to describe **a fine person, a sweetheart**, perhaps in the same way that **my bullies**, or **my bully-boy**, was also used about one's mates or a lover. Women still sometimes say, "I like a bit of rough" when they refer slightly disparagingly to their slightly downmarket partners. Could there possibly have been a spelling change ages ago that we've forgotten?

Of course, a RUFFIAN might actually wear a ruff, as shown in the phrase from 1623, "to lag like a Ruffian's starcht Ruff in a storme of raine." We know that the ruff as fashion developed from around 1555 as a *fraise,* a narrow ruched band applied as a final strip or accent at the top of a bodice. This style probably started in Italy and spread throughout Europe.

Collars starched with rice water were worn in Ceylon and India to keep men's long, oiled hair from soiling their robes, and Western traders imported this idea. Starching allowed greatly enlarged and more upright, stiffened ruffs. These first came to England via the Low Countries, and Britain always lagged behind trends in Spain, France and Italy. Each country favoured different styles of **ruff**, but all of them got bigger and more extravagant over time.

Governments everywhere passed disapproving sumptuary laws from time to time in efforts – generally in vain – to curb the more absurd excesses. A large, pleated ruff could require a couple of yards of lace, and its many tiers were held up by an elaborate, wired frame. Parisians used to shout after courtiers so adorned, "You can tell the calf's head by the ruff!" meaning that they looked stupid, like a boiled calf's head on a plate dressed with a paper frill.

By the close of the 16th century, near the end of Elizabeth's reign in England, the mode had veered towards the 'falling collar', a much simpler style. However, pleated ruffs kept going in and out of fashion for at least another 40 years. Clothing evolved gradually during the 16th and 17th centuries, with innovations arriving in cities and ports first, then spreading outward. In the deepest countryside, costumes could be a generation behind the latest craze in London. Any rural folk arriving in the Big City could easily be identified as yokels from their dress. Portraits show a huge variety of ruffs and collars for both the aristocracy and gentry over this period, and there were lots of published comments about those who had fallen far behind the 'cutting edge'. Mocking other people's manner of clothing and adornment was an easy means of feeling superior – and if you read today's so-called 'popular Press' or Twitter, it still is!

We don't know how poverty-stricken peasants dressed, because they were rarely represented in art. Painters have traditionally relied on wealthy patrons, who were only interested in depictions of rich clothing. However, in the late 1550's, post-Reformation England was in a serious financial situation, subject to the dislocation of many people, and economic 'boom and bust' – rather like our own time. Quite a few of the aristocracy were ruined; previously landed and well-born families were reduced to penury.

Over the next 50 years, rural areas were full of men and women living outside society, many lacking the skills of craftsmen, or any practical means of supporting themselves legitimately. The men of these outcast groups might well cling to the vestiges of old clothing from a better life, wearing ruffs when these had long gone out of fashion and been replaced by plain white bands of linen. Or they might sport stolen finery inappropriate for their low status. So, if you spotted a **bandless man**, wearing an obsolete and tattered ruff, he would almost certainly be **a violent robber**.

I would suggest that this secondary meaning of RUFFIAN arose from the many such marginalised folk living perilously on the edges of society: soldiers turned off without pay; escaped prisoners, the landless poor, the mentally ill, the learning disabled, the old and infirm, abandoned children, widows and orphans. Such vagrants were regarded with suspicion and fear and at the time when there was no social security provision, what could they do? Staying 'honest' under these circumstances was rarely possible; begging, stealing and prostitution were rife. This parallels the downfall of so many LOOSE WOMEN. We are not condoning it, but recognising a tragic reality.

Finally, a modern use of the root word is RUFFING – described by the Urban dictionary as a polite **form of applause**, usually for an after-dinner speech in a university or similar venue, in which **the audience stamp their feet or beat their hands** on the table to show appreciation.

RUG ☞ RAG

A mysterious Old Norse word *roggvathr* meaning 'tufted' was worn down to a rag, through *rog*, a tuft of hair. This gave us the word rug as well as rag, though **rug** took a while to shed the sense of hairiness and apply only to a household floor covering – maybe because early rugs were likely to have been animal skins anyway.

I hope no one will PULL THE RUG FROM UNDER ME, as it would be painful. It is not very likely that I would really stand still and let you do it, so the idea probably comes from theatrical slapstick, and was never meant to be taken literally. This is quite a recent figure of speech, probably only coined in the 20th century for **anything that sabotages your plans by withdrawing support**.

However, the saying SNUG AS A BUG IN A RUG sounds **nice and comfortable**, particularly when tucking up a child in bed, and the rhyming repetition no doubt helped people remember it. Although it sounds quite modern, similar phrases were known in the 17th century, such as "He sits as snug as a bee in a box". In an anonymous work, which appeared in David Garrick's 1769 Shakespeare Festival publication, we find, "If she [a rich widow] has the mopus [money] I'll have her, as snug as a bug in a rug".

At the other end of the scale from rich widows, we have the term RUG RATS for **babies at the crawling stage**, an evocative term, used in an Australian cartoon series about infants and their wild and devious stratagems.

By the way, the word **rugged**, as in "he was handsome and as rugged as a lumberjack", came from the same shaggy origin as rug and rag, and only lost its hirsute implications later. Nowadays, you might be a rugged individualist meaning **stalwart and staunch**, or speak of climbing in the rugged hills suggesting they are **craggy and a bit steep**, yet see never a hair. Well, that's the way words change. You know what they say: "hair today, gone tomorrow".

{See also my book *Rigmaroles and Ragamuffins* under RAG}

SABOTAGE ☞ see CLOGS

A SABOT is French for the wooden clogs the peasants wore. I remember when quite young being charmed by my father's story that the word SABOTAGE, meaning **malicious damage to your employer's property**, came from French factory workers **throwing their sabots** into the machinery to express their displeasure with the bosses.

I now find that the story was not quite true and is something of an urban myth but still near enough. A *sabot* is a wooden clog and *saboter* in French is **to walk noisily and clumsily**. This lent itself to the idea of **clumsy work**, and then of deliberately **spoiling what you were doing as a protest**. The word for this, SABOTAGE, has been used in France for centuries, but has only taken root in English during the last hundred years.

SACK ☞ BAG

A Latin word for a bag, *saccellus*, had a diminuitive, *saccus*, which became Old French *sachel* and so became our English satchel. A good, strong leather bag is handy for books, and **satchel** was applied from the 16th century to the common school bag – and, of course, also gave us SACK. To SACK A TOWN carried the idea of **plunder thrust into a bag**.

TO BE GIVEN THE SACK has meant **getting rid of a workman** since the 16th century. Many workers traditionally owned the tools of their trade and carried them from job to job in a sack. When they were dismissed, the boss would give them the sack so they could gather up their tools and hit the road. At least this meant they could go and work elsewhere, but if they had done some really awful work or broken too many rules, their tools were not given back but burned, in other words they were literally 'fired'. Well, that's the story anyway.

A SAD SACK is **an ineffectual person**, said to derive from Private Sad Sack, a hapless cartoon character in the World War II era American comic book. Common military slang for an inept person was, "a sad sack of s**t", and the artist, Sergeant George Baker, simply shortened this. Historically, it was not uncommon to refer to clothes that were limp and worn as **sad**, and a *sack* or *sacque* was a type of loose gown in the 18th century, originally informal dress, which became increasingly ornate and formal by the end of that century. There was a measure of disapproval for them, as some aristocratic ladies wore these loose garments to conceal their illegitimate pregnancies – back to our loose women again.

SACKCLOTH AND ASHES

Only one step up from nakedness was TO WEAR SACKCLOTH AND ASHES, from a tradition in classical times of stripping off all normal attire, smearing yourself in ashes or dirt, and wearing only a piece of rough cloth – usually made from scratchy, black goat hair – to show such emotions as remorse, despair, grief or shame. Today, you would use the term to show **you are truly sorry**, or **you are mourning**, without necessarily emptying the ashtray over your head.

SHEEPSKIN

In everyday use, a SHEEPSKIN is the fleece or the whole skin of a sheep that can be worn as a cloak, made into a coat or boots or even a nice rug for the floor. However, with the wool taken off, the bare skin can be scraped and used as parchment for writing on – particularly authoritative documents, such as university diplomas. A SHEEPSKIN can mean not only **the official scroll** showing which degree was awarded, but also the actual **person** who gained this qualification. According to modern urban slang, a SHEEPSKIN is **an arrogant young man who makes a pass at your girlfriend!**

Perhaps he's been making SHEEP'S EYES at her for ages, which means **to look at someone longingly** or **amorously**. To say someone is or looks SHEEPISH is to assign them attributes often linked with sheep, like being **simple minded, excessively meek and submissive**. We do rather take it for granted that sheep are very benign – though if you walk across a field of Swaledale ewes at lambing time you might think differently. They have fearsomely curly horns and won't hesitate to use them. The Moroccans pin it down in a proverb, 'IF WE WERE AFRAID OF SHEEP WE WOULD NOT WEAR WOOL.' Maybe they should come and see the Ewes in Swaledale for themselves – "be afraid, be very afraid!"

{*For more interesting sheepy information see my book* **Rigmaroles and Ragamuffins** *under WOOL*}

SHEET ☞ BLANKET, PALL and SHROUD

The common origin of this word, used in so many ways, was a probable Germanic word *skaut* or *skit*. As it developed in English it threw off shoot, shout, shot, shut, skit, Scot-free, and to scuttle a ship. Not exhausted by all this activity the word gave rise to English *scete*, meaning 'cloth', and *sceata*, meaning 'a sail-rope'. These two words coalesced in form, but kept the two separate uses of SHEET. A busy, busy word for all those activities, followed by a nice nap UNDER THE SHEETS meaning **in bed**. Once we are BETWEEN THE SHEETS we are also **in bed**, but the word has more than one meaning. If we get them confused we might find ourselves trying to sleep half way up the mast of a ship, since SHEET not only means **bed linen**, and **a piece of fabric**, but also the **ropes on a sailing ship**. Also, A SHEET ANCHOR is the strongest, main anchor of a ship, and is used metaphorically to mean something or someone **totally reliable and dependable**.

We use the term A SHEET OF SOMETHING, such as snow or sand or for any broad expanse or something stretched out, such as A SHEET OF PAPER, and these phrases go back to the 16th century. Later on we got A SHEET OF FALLING RAIN, and SHEET LIGHTNING which is lightning reflected onto the clouds in a broad band.

A **newspape**r as a SHEET or a BROADSHEET, and SHEET-MUSIC for the words and musical notation of songs, are quite recent terms, and we say WHITE AS A SHEET as a simile for **the whiteness** of linen or cotton when washed and bleached.

A WINDING SHEET takes us back to SHROUD and PALL, as it is also the cloth used to wind around and **cover the dead**.

SHIRT

So now we come to men's wear. First, let us slip into the SHIRT. As we saw, shirt and skirt started from the same root, probably meaning short. Later, the words took their separate ways. When you call someone shirty, you mean **irritated, short-tempered** or even **angry**. When we say, KEEP YOUR SHIRT ON, it is – as with suggestions to keep on other items of clothing – an exhortation to stay cool, chillax. In the 19th century, the opposite idea, now obsolete, was GET YOUR SHIRT OUT, or HAVE YOUR SHIRT OUT, possibly suggesting **the dishevelment caused by anger** and. therefore, chivvying you **to get your shirt and coat off in order to be ready to fight**.

··· *Shirt* ···

As we saw in a previous entry, one of the impossible tasks given to the girl in the folk song Scarborough Fair, with its magical herbs, is the instruction by the young man to a third party,

> "Tell her to make me a cambric shirt,
> Parsley, Sage, Rosemary and Thyme,
> Without any stitch, nor needle work,
> Then she'll be my true love again."

TO PUT YOUR SHIRT ON A HORSE is **to bet on it**, with the implication that you stake all you have. Thus, if you lose, you will have NO SHIRT TO YOUR BACK, which means **owning nothing whatsoever**. TO GIVE SOMEONE THE SHIRT OFF YOUR BACK means **to be very generous** and has been around since the 18th century.

In medieval times, if you committed a lot of sins, you might repent of them and decide to wear a HAIR SHIRT in order **to remind yourself to behave** and **self-punish or mortify the flesh** – which, I have to say, is not my idea of sin and forgiveness! The garment was otherwise known as a *cilice,* originally fashioned of rough goat's skin. A HAIR SHIRT was worn, of course, with the prickly fur inside, often augmented with twigs and wire to irritate more. Since it was also considered unsuitable to wash yourself for long periods while doing penance, the united pong of goat and human must have been as much a penance for anyone in the same room as for the sinner.

The great French playwright, Molière, wrote a farce about a chap called Tartuffe, who is exposed as a religious hypocrite when he is found to be wearing his hair shirt the wrong way round, with the uncomfortable bits safely away from his skin. One can only admire his cheek!

Modern urban slang uses the term A SHIRT for a man **whose idea of a good time is a night out clubbing, making advances to women, and getting very drunk** – not necessarily in that order – finishing off with an inebriated brawl and a police cell, probably. Quite a long way from the idea of doing penance. A STUFFED SHIRT, on the other hand, is **a pompous, probably rather stupid man**. I'm not sure I see much difference between the two – and both are to be avoided.

The shirttail is the lower part of the shirt, hanging below the waist and worn either inside or outside the trousers, depending on how casual a fellow is. Modern slang gives us SHIRTTAIL as **mean** or **paltry**. A SHIRTTAIL TIME therefore can describe being poor, your shirt hanging out, and hence a figure of speech for **a miserable time**, whether an evening or a lifestyle, as in WE WERE SHIRTTAIL POOR. However, SHIRTTAIL CHILDREN or SHIRTTAIL KIDS simply means very young ones who cling to one's shirttail as they toddle about.

Shirts can be used in sinister magic, as in the myth of how the Greek hero, Heracles (Hercules to the Romans), met his end. The dying Centaur, Nessus, tricked the hero's wife into giving him a poisoned shirt to wear. So there are times when you are better off undressed.

That reminds me of the story of a King who became ill, and being a little irritable as a result, kept cutting off the heads of those who failed to cure him. At last, he was told he would be well again if he could just sleep one night in the shirt of a happy man. His advisers searched the country, but found no one who was truly happy, because the country was riddled with injustice and tyranny, and they took back all these tales of wrongdoing and misery to the king. He sent them out again, fortunately still with their heads, to try one more time to find a happy person. Eventually, they found a beggar lying on his back in the sun, laughing. When asked, he told them, "yes, I am truly a happy man", but when they offered him a hundred gold pieces for his shirt he laughed even more and told them he hadn't got one. The courtiers went back and reported to their monarch, really fearful of losing their lives this time. But in their absence, the King had become ashamed of his people's suffering and had begun to work to put things right. As their wellbeing increased, the ruler also became healthy and content, and stopped chopping off people's heads. Result!

SHOES ☞ BOOTS

From earliest times, mankind has sought to protect their rather tender tootsies from thorns, sharp rocks and bad weather. Early shoes were probably strips of leather or cloth wound round the foot and lower leg, and remnants of these are sometimes found as artefacts in prehistoric tombs, or wall paintings. The word SHOE probably comes from the Indo-European base *skeu,* meaning to cover. This root yielded up the ancient Germanic *skokhaz,* and a lot of words similar to shoe in European languages – but none elsewhere. Many appear akin to the source and offshoot leading to stocking.

Shoes have huge significance in superstitions and folklore, far too much to more than hint at it here. We are looking at their impact on our language, but if we don't take a moment to look at this rich storytelling background, we won't understand why we have many wonderful phrases relating to boots and shoes.

Both shoes and feet can symbolise the whole person. For example, shoes have been used in the past as an addition to, or perhaps a substitute for, human sacrifice. It is not unusual to find footwear, particularly children's shoes, built into the foundations of old buildings. Shoes and boots are both reverenced and despised; they are lowly, and defiled by dirt, yet they were traditionally precious possessions. They may speak plainly of the wearer's status, or peek out coyly from under a woman's dress. Thus, shoes create some of the same ambivalent feelings as body openings, because we slip our feet into their receptive interiors. Indeed, by suggesting the sexual organs, the source of life itself, they carry a powerful emotional charge and create erotic feelings inmany more people than the tiny minority with an admitted foot or shoe fetish.

We still hang shoes on the back of the wedding car to promote fertility, and you see them on the backs of lorries in India, along with ornamental hair tassels, as potent guarantees of good luck and safety. If you've seen the way people career around honking horns madly in India, drivers need all the good luck tokens possible – as do pedestrians!

It is quite normal for children to be obsessed with shoes, and when they are bought new ones they will show them off to everybody they meet. Naturally, therefore, there are plenty of mentions of shoes in Nursery rhymes.

There are requests:

> "Cobbler cobbler mend my shoe,
> Get it done by half past two!"

There are laments for their loss:

> "Cock a doodle doo!
> My Dame has lost her shoe,
> My master's lost his fiddlestick,
> And knows not what to do."

There's even an old woman who lived in a shoe. And, of course, there are the counting rhymes such as:

> "One, two buckle my shoe"

There are many other shoes and boots in fairytales, such as Seven-League boots, a story about magic boots that can travel great distances. Another neat device is to have a character who wears out many pairs of shoes made of iron on his or her quest, to suggest the supernatural distance covered.

There is also a tale in which the devil decides to go and torment a town – usually named as one near where the story is being told – and asks a man he meets how far he will have to walk. The man happens to be a quick-witted cobbler with a sack of old shoes to mend, and he shows them to the devil saying he has worn them out on his journey from the town. The devil is discouraged by the distance implied and goes away leaving the neighbourhood untormented.

SHOON used to be the plural of shoe, and we find this form occasionally in folk rhymes or poetry, as in Walter de la Mare's verse:

> "Slowly, silently now the moon
> Walks the night in her silver shoon".

If you threaten to CAST ME OFF LIKE AN OLD SHOE, I might be scared enough to SHAKE IN MY SHOES, since the first means you are going **to get rid of me as a friend**, or cut me out of your life and the second that I am **trembling with fear**.

TO STEP INTO SOMEONE'S SHOES is **to follow them** into a job or perhaps **an inheritance**, made more explicit in the term, WAITING FOR DEAD MEN'S SHOES.

It is a good thing to try and PUT YOURSELF IN SOMEONE ELSE'S SHOES or WALK A MILE IN SOMEONE ELSE'S SHOES so as **to understand their problems**, best known in the Native American proverb, DO NOT JUDGE A MAN TILL YOU HAVE WALKED A MILE IN HIS MOCCASINS.

A more cynical take on this goes, "TRY AND WALK A MILE IN ANOTHER MAN'S SHOES. THAT WAY YOU'VE GOT HIS SHOES AND A MILE'S START ON HIM".

You might say YOUR HEART IS IN YOUR SHOES – as with BOOTS, when **you feel low and unhappy** about something. Another saying is, TO FEEL WHERE THE SHOE PINCHES. Just as you can only know where a shoe doesn't fit by wearing it, **some problems you only discover by trying the thing out**.

A colloquial saying that emphasises **a contrast or disappointed expectation** could be on the lines of, "We gave that friend all the help we could when she had problems, but when we asked for her help, THAT WAS QUITE ANOTHER PAIR OF SHOES." In other words, the friend had refused to help in her turn.

If we want to describe a girl as **rather too good to be true**, we might sneeringly say SHE'S A LITTLE GOODY TWO SHOES, or A GOODY GOODY, which comes from a book of 1765, thought to have been written by Oliver Goldsmith and called *The History of Little Goody Two-Shoes*. It is unlikely that many people nowadays have read or even heard of the book, and the phrase probably became popular because of the stage pantomime LITTLE GOODY TWO-SHOES which has a hotchpotch of nursery rhyme characters, and no discernible plot.

There is a classical version of Cinderella from ancient Greece in which the heroine, Rhodope, finds a Lydian shoe dropped by an eagle. Ancient Lydia was famous for its shoe-making industry, which created beautiful, aromatic leather footwear.

In ancient Rome, it was considered an ill omen if mice nibbled your shoes. A superstitious man asked Cato, the famous philosopher, what he would think if this happened to him, and he replied that there would be nothing extraordinary in it, "but it would have been really wonderful if my shoes had nibbled the mice" – perhaps the earliest version of the 'man bites dog' plot.

In Britain, we have a proverb that says, THE SHOEMAKER'S CHILDREN ARE THE WORST SHOD whose sentiment is probably echoed in most cultures, the Arabic version being, THE COBBLER MUST GO BAREFOOT, THE WEAVER MUST GO NAKED, meaning of course that those who do a particular form of work are less likely, by inclination or time, **to provide their skill to themselves or their families**.

There is a Norse hero and minor deity called Vidar, whose time was entirely devoted to collecting waste pieces of leather from shoemaking throughout the ages. With these scraps, when the world comes to its end during Ragnarok, the Twilight of the Gods, he will make a great shoe and use it to tread on the lower jaw of Fenrir, the magic wolf who killed his father Odin. And then he will rip the creature's jaws apart to kill it. You can help in this heroic task by throwing waste pieces of shoe leather away where he can find them. It sounds like the ultimate in recycling.

If I say THE SHOE or THE BOOT'S ON THE OTHER FOOT, I mean that a situation has changed and **is now more favourable for me than for you**. It dates as a saying from the early 19th century, when lasts or wooden forms on which footwear was fashioned were the same shape for both right and left feet.

To describe someone as DOWN AT HEEL suggests that they haven't the money or perhaps inclination to replace their worn-down shoes, just as does SLIPSHOD for someone in ill-fitting shoes – figuratively, both mean **slovenly and careless**. WELL-HEELED or WELL SHOD suggests **a wealthy individual**, and TO COME INTO THE WORLD HOSED AND SHOD is **to be born into a good estate**.

A SHOE may be fastened in many different ways, such as laces, string or twine. LIVING ON A SHOESTRING is rather a poetic way of implying that someone **has very little money and is having to manage on a budget stretched tight and thin**. SHOESTRING POTATOES are cut very thin and fried crispy, while in 1880's America, there was a disreputable group of small-time crooks called SHOESTRING GAMBLERS, who only played for small stakes, and therefore preyed mainly on poor people. A shoestring is essentially the same as a shoelace, and you can still buy the sweets known as LIQUORICE SHOELACES.

In the Bible, we find reference to a shoelatchet – a leather thong used like our shoestrings to tie or lace up ancient footwear.

> "And Abram said to the king of Sodom, I have lift up mine hand unto the LORD, the most high God, the possessor of heaven and earth, That I will not take from a thread even to a shoelatchet, and that I will not take any thing that is thine, lest thou shouldest say, I have made Abram rich."
>
> [Genesis 14:22–23, *King James Bible*, 1611]

It is said that, after his death, Saint Cuthbert's shoes continued the healing work of his exemplary life. When they were put onto the feet of a sick, lame man, he became well and finally skipped, danced and ran off rejoicing. Perhaps his were the original 'Happy Feet'.

A DEALER'S SHOE refers to a modern device in casinos and other gambling dens to keep several decks of cards in, so as to speed up the game and prevent sly cheating by slipping cards in and out of a pack.

SHROUD

NO POCKETS IN A SHROUD! is one of the commonest phrases quoted to me when I have occasion to mention the word – not that this is often, thank goodness.

The word came originally from a prehistoric Germanic base which spread and evolved throughout Europe. It became *skreud, skruo, skreuo, scraud* and *scrud* – all from a verb meaning 'to cut' and hence, eventually, a noun, 'a cut piece.' This later led to both SHRED and SCROLL. An Old Norse derivative, *skruoi*, was used for any sort of fittings, or textiles.

Some dialect words to do with pruning a tree or bush come directly from that early connection with cutting or lopping and terms for bits of cut off things, such a nail clippings, so that **the trimmings** from a bush or a plant may be known as SHROUD.

SHROUDS, of course, are the cloths traditionally wrapped reverentially around the bodies of the dead, and can be of many different types. In the 1500's, SHROUD described any simple garment, but by the turn of the century, it had developed into to a white cloth on which to lay out and envelop a corpse. Finally, it settled into the modern sense of a garment used exclusively for the dead, as a winding sheet.

From the basic idea of a body wrapper, we get many uses of the word to imply things **that cover or even shelter us**, such as SHROUDED FROM PUBLIC

EYE. You might have A SHROUD OF RAIN on **a wet day**, or be SHROUDED IN GLOOM if you are feeling almost deathly **miserable**.

The SHROUDS on a sailing ship are **a set of paired ropes** attached from the top or part way down the mast, used as part of the standard rigging to stabilise the mast. These ropes or cables are attached to chain plates attached to the hull, and take some of the strain in a high wind.

SKIRT ☞ SHIRT

It's not clear why SKIRT, essentially the same word as SHIRT, has ended up as a woman's garment hanging from the waist, while modern Icelandic *skyrta* is a sort of long shirt with full tails hanging well below the waist. Related words in other European languages may have been borrowed from Middle Low German, the *scorte*, an apron.

Calling a woman A BIT OF SKIRT or even A NICE BIT OF SKIRT is a way of referring to a woman by **what she wears** rather than as an individual and is probably a veiled allusion to the genital area. Not really appropriate today anyway, when so many women wear trousers.

A SKIRT may be long or short, and unless stylishly cut on the bias or made with triangular gores for swishy movement, it has often been a rather boring garment that simply covered whatever was considered 'naughty bits' at any particular date. The SKIRT of anything can be **the margin, edge, verge, border, rim,** or **outer portion**. The **outer areas** of a town are called its OUTSKIRTS, which is quite an odd idea, if you try to picture it.

Even a saddle has SKIRTS; they are **the leather flaps on either side extending down the horse's flanks** for a more comfortable, chafe-free ride.

To SKIRT AROUND a topic or a problem is **to avoid it**, to keep it at a distance as it were, and the SKIRTING BOARD in a room is **the wooden surround that runs around the bottom edge.**

Here's a rhyme from Australia, where a 'cocky' is a dairy farmer as well as a parrot:

> "A cocky's daughter from Sale
> Made a skirt of hay from a bale,
> But a marauding young calf
> Ate her school skirt in half,
> And the rest blew away in a gale."

SLEEVE ☞ SHIRT

The sleeve almost has a life of its own in our language, with parallels in folklore, where a part of a garment may stand not only for the whole, but for the wearer, too. Long ago, having your baby snatched away by supernatural or evil beings and replaced by a changeling must have been a great worry, considering the number and variety of preventive measures recorded. One such measure was to lay the sleeve of the father's shirt across the cradle. It seems obvious that this represented the father's strong protective arm covering the baby, even in his absence during work or sleep.

The word SLEEVE came to us probably from the rather nice sounding Middle Dutch *sloof* or *sloove*, and in Middle English it became *slif, sliefe, sluf*, or the verb *slupan* – all having the meaning of to slip or glide. In this case, probably it suggests the action of sliding an arm into a sleeve, just as with SLIP for a garment – 'something you slip into'. Indeed, the action verb may have generated the noun as 'that into which the arm glides'.

You might want to KEEP SOMETHING UP YOUR SLEEVE, because it means **to hold something in reserve, keeping it a secret**. That's almost the reverse of WEARING YOUR HEART ON YOUR SLEEVE, because that denotes **making it very obvious to everyone that you are in love;** this sprang from the practice of young men in medieval times of wearing a "favour" or coloured scarf from their sweetheart fastened on their sleeve, especially when jousting in a tournament.

In a more modern way, we slip an audio or video disc into its **sleeve – a cardboard or plastic cover** that protects it, just as we protect our arms with textile sleeves.

> Question: "Napoleon, where do you keep your army?"
> Answer "Why, up my SLEEVY!"
> [Well, there aren't many textile jokes. Sorry. It's just awful!]

TO HANG ON SOMEONE'S SLEEVE is **to flatter them with lots of attention**. Perhaps it came from a time when men wore very long elaborate sleeves and you could literally hold onto them. In the days when sleeves were wide enough to hide one's face, someone deriding you might very well LAUGH UP HIS SLEEVE, really only slightly concealing his mirth. One sees films of Japanese geisha girls doing the same thing behind their broad kimono sleeves. This English phrase is first recorded in 1546 and may have originated with a 16th century proverb; it implies **you are being mocked secretly**.

TO ROLL UP YOUR SLEEVES is **to get down to work**, not necessarily to do a practical job any longer but definitely **with a sense of purpose and determination**.

Being without sleeves used to be a figure of speech for **uselessness**. SLEEVELESS, and similarly BOOTLESS, used to mean **futile or feeble activity**, leading to nothing, but both are now archaic. Presumably people really low down in society would not have proper shirts or footwear, and might be despised for this. To go on A SLEEVELESS ERRAND was a figure of speech for **a pointless exercise**.

Alexander the Great introduced his fellow Greek men to the Persian fashion of adding long sleeves to their tunics in the 4th century BC. However, for hundreds of years it was decried as decadent, dangerously effeminate, and liable to hasten the end of the Empire.

One of the more gruesome Viking tales, the Völsunga Saga, concerns shirtsleeves. A Viking woman called Signy had two sons by her husband Siggeir. She planned to send them to serve her brother Sigmund as warriors. To test their valour, she sewed the sleeves of their shirts to their hands. Unfortunately, they screamed with pain, thereby failing her test; so, on their mother's instructions, as soon as they arrived, Sigmund killed them. Later on, Signy had another son, Sinfjotli, by an incestuous union with Sigmund. By the time the boy was 10 years old, he had grown very tall and strong. He did not flinch when his mother sewed his shirt sleeves to the backs of his hands, nor even when she went on to tear the shirt off, skin and all, "Such pain is little to a Volsung", said Sinfjotli, and went off to join his father/uncle.

I know mothers worry about garments getting lost by the local laundry, and often embroider their sons' initials inside clothing when they are sent off to boarding school or college. Even so, I think Signy was a teeny bit OTT, don't you?

SMOCK

The Old English word *smoc* gave us the term SMOCK for a loose-fitting overshirt or gown for men or women, particularly agricultural workers. The name may also go back to words for the action of pulling on a garment over the head.

Many SMOCKS were decorated by rows of little pleats at the upper front chest, upper arms and cuffs. These helped to provide a better fit and gave us the term SMOCKING for the **decorative embroidery** stitched over the gathers.

Alas! For some reason the word SMOCK has aroused the ire and even worse instincts of the young men of today and there are a number of modern, non-textile uses of the term – of which the only repeatable one is to describe a man with a very small penis. It can also mean **a sharp blow on a sensitive portion of the anatomy**.

SOCKS ☞ SHOES, STOCKINGS

After all that excitement, SOCKS are a bit of a let-down. The original Greek word for a foot covering or light shoe was *sukkhos*, perhaps borrowed from an Asian language. The Latin *soccus* became *sok* in prehistoric German.

BLESS YOUR COTTON SOCKS or even BLESS YOUR LITTLE COTTON SOCKS! is **an affectionate exclamation** of little real meaning. BOBBY SOXERS were 1940's US **girls in ankle socks** who loved pop music and performers like Frank Sinatra. They probably formed America's first distinctively teenage culture – although, in practice, the group members' ages extended to their mid-twenties.

The origin of the term TO SOCK, meaning **to hit**, is not known, and may not have anything to do with socks as such, but nowadays does tend to be thought of as a blow with a viciously weighted sock. It was well known as meaning **to hit or strike** since the early 17th century.

SOCK IT TO ME is a colloquial way of saying, **"tell me"** or more particularly, **"amaze and shock me!"** Telling someone to PUT A SOCK IN IT is said to come from the early days of the gramophone which is itself now virtually obsolete. The original gramophones had a large, flaring horn to magnify the sound of the record, and no volume control, so you literally had **to stuff a sock into the horn to quieten it**. We use the phrase impolitely to say, **"Shut up!"** I suppose with those old fashioned gramophones you could simply close the doors at the front of the cabinet to dampen the sound also. Not as much fun as shouting PUT A SOCK IN IT.

If I were to suggest to you that we might play with a SOCK PUPPET, you would probably be pleased (if, like me, you enjoy the simple pleasures in life). The term used to refer solely to those endearing **soft hand puppets** made from odd socks orphaned after emptying the washing machine. [First Rule of Life: There is always at least one odd sock left over from a big wash.]

No longer! A more sinister type of SOCK PUPPET has come into being in recent years of these internet savvy times. It is **a fake identity assumed online**. These IDs are assumed to dupe people out of money, to conceal one's name when doing online dating, or even to pursue children online. They are also used by firms to fake good product reviews, or by authors trying to promote sales of their books with pretend "couldn't put it down!" comments.

Come on socks! If you are to shed such evil alter egos and join the fashion world, you will have to PULL YOUR SOCKS UP! meaning **you must make more of an effort.** The unglamourous sock was probably best summed up by a character in a play by Alan Bennett, from the 1970's: "One of the few things I have learned in life is that there is invariably something odd about women who wear ankle socks." [Oh dear. Guilty as charged.]

SPORRAN ☞ PURSE

That interesting article of dress worn on a chain over the abdominal area of a Scotsman's kilt is called the SPORRAN, and this furry or tasselled purse is where he keeps his [jealously guarded] money. The word PURSE came from the Latin, *burse,* but was borrowed by the Celts, probably with the money still in it [Hey! Give that back, you no-good Celtic *****] where it became the Gaelic *sporan* and eventually SPORRAN.

Nowadays, they are infrequently worn, except with full highland dress. It can best be described as a hairy purse. I regret to say that most of the figurative uses of SPORRAN relate – you've guessed it – to the male scrotum and to the male ego of a certain type of rough Celt. A SPORRAN SLAP is used to describe **slapping a woman across the face** when he feels she has disrespected him. A woman who refuses his advances may escape a slapping, but be disparaged as **an aggressive lesbian** or SPORRAN BASHER.

Let me put the record straight. I love Scotsmen, who are a wonderful, handsome, generally kind race of men to die for, and I would love to go out with one on a Saturday night for a wee dram. I'll even pay my share, so there's less need to dip into that splendid, ornate sporran of which you are obviously so proud.

STOCKINGS ☞ FROCK, SHOES and SOCKS

STOCKINGS have been around for something like 3,000 years, since the craft of knitting on simple, circular, pegged frames arose in the Middle East; that was even before the ancient Egyptians depicted this craft in their tomb paintings. However, in the West, coverings for legs and feet were fashioned from cloth until the comparatively late luxury of hand-knitted stockings.

Nowadays they are used to describe the leg coverings women wear, typically made of translucent nylon or silk, fitting closely and often held up by suspenders or elasticated garters. In earlier times they were a similar tubular garment for men, more commonly of wool and knitted.

Here's a nice traditional knitting rhyme from a record of country lore of 1890:

> "Needle to needle and stitch to stitch,
> pull the old woman out of the ditch.
> If you ain't out by the time I'm in,
> I'll rap your knuckles with my knitting pin."

You 'pull the old woman out of the ditch' by the wrapping and looping action to create the knitting stitch. It must have helped teach children doing plain and purl. Other rhymes were used for counting stitches or counting rows. In past times, when many if not most country women were illiterate and knitting was a family or communal activity, such little mnemonics must have been both useful and fun, bonding the generations.

It seems the word STOCKING probably came in during the 15th century, from the wooden **stocks** in which one's legs were imprisoned for misdemeanours – typical black humour of the time. Or it may be derived from Old English *stocc* meaning trunk or log, because swaddled or covered legs looked a bit like tree stems.

In the 16th and early 17th centuries, the fashionable male image was a solid square body on slender legs with well-turned calves and good ankles – think of cartoons of Henry VIII. Men tied their woollen hose to their doublets by strings – a sort of early suspender. About this time, knitted hose replaced tailored cloth ones – particularly for the luxury market. Queen Elizabeth I was passionately devoted to knitted silk stockings.

The first knitting machine was invented in the 16th century to replace laborious hand knitting. Alas for its inventor, William Lee Vicar of Galveston, Nottinghamshire, Queen Elizabeth refused to support his innovation, and he died in France in penury. Ironically, the French eventually made a great success of the machines he left behind, and his brother exported the industry back to England. However, it was more than a century before machine knitting really took off.

Stockings were a popular theme in songs and poems, as in this Ring Dance:

> "Red stockings, blue stockings,
> Shoes tied up with silver,
> A red rosette upon my breast
> And a gold ring on my finger.
> So here we go around, around,
> And here we go around;
> Here we go around, around,
> Till our skirts shall touch the ground."

Apart from a BLUESTOCKING [qv], and in spite of the many references to them in folklore, stockings have failed to impress themselves on our language through figures of speech. As with other garments, in traditional stories or beliefs, stockings often **symbolise the whole person** and occur in a variety of rituals. They can be used in healing, for magical protection and to attract a lover in ways other than the obvious one of seducing him by showing one's stockings right up to the garter – or by removing them tantalisingly. A girl might put out a stocking at night and recite **a spell to draw her lover to her**.

In many parts of the world, women's stockings display special colours and patterns reserved for specific occasions, particularly marriage ceremonies; here their removal is part of the bride's undressing – in a token way publicly, or completely in private – by the groom. Stockings could be symbolically crossed under beds, hung on walls, and thrown after the married couple as we now throw old shoes. The bride was supposed to place her stockings crosswise on the bed on the first night or she would not have children. Your left stocking tied round your neck supposedly cured a sore throat. And let us not forget the importance in our Western culture of the STOCKING FILLER at Christmas – **small gifts** that may be put in the CHRISTMAS STOCKING [qv].

It seems that, in the 18th century, stockings themselves held no erotic appeal, which was reserved for the GARTER [qv]. Following the domestic respectability of the mid 19th century when all below the waist was unmentionable there came an intoxicating time when women's feet and ankles were on show again under the crinoline, and stockings were once again embroidered and decorated.

Apparently in the early days of STOCKINGS, those annoying 'ladders' that run when you snag them and pull a stitch, were known in the 18th century, as 'Louse Ladders'!

> In olden days a glimpse of stocking
> Was looked on as something shocking.
> But now, God knows,
> Anything goes.

Cole Porter wrote that in the 1930's, and it has lost none of its relevance. In our 21st century, when even more obviously 'anything goes', one might have supposed that hosiery and underwear generally would be less intriguing and titillating – but, of course, this hasn't happened. Why? Perhaps too much nakedness becomes boring; partial undress and rare glimpses of what is normally covered by clothing have always been more erotically arousing than total nudity.

In our more 'open' society, even children are able to share things that previous generations would have kept from their mums and dads, like this rather delightful Australian combination of Cole Porter and a nursery rhyme;

> "Shocking, shocking, shocking,
> A mouse ran up me stocking.
> He got to me knee
> And what did he see?
> Shocking, shocking, shocking".

STRAITJACKET ☞ MEN IN WHITE COATS

We use the term A STRAITJACKET figuratively to allude to **anything that restricts you against your inclination**, such as rules and regulations, or conventions that might irk you. The straitjacket was a tough canvas body wrapping, usually with very elongated sleeves and straps that could be tied right round the body completely immobilising the arms. Once they were standard issue in prisons and lunatic asylums to confine violent individuals. My mother, who worked in the UK's healthcare field in the 1920's and 30's, was familiar with their use. Today, they are only to be found in hospital and prison museums. Attitude changes and improved medication eventually did away with the necessity for such physical restraints, though **the drugs** in use now are sometimes referred to as a 'chemical cosh' or a CHEMICAL STRAITJACKET.

SUBTLE ☞ VEIL

To say something is SUBTLE is to suggest it is **indirect** or **difficult to see**, something **clever, delicate, elusive, not obvious**. That is exactly right, because it drifted down through the millennia from a Latin word, *subtilis*, which described a fine, delicate cloth. Used particularly for veils, it was so gossamer thin as to be almost transparent, and could be seen through by the wearer. The word was derived from *sub tela,* a web, which itself came from *texare,* to weave, hence literally means 'underwoven.'

In medieval France, the word became *soutil,* then crossed the Channel into Britain with the Normans, as *sotil.* Fussy pedants in the 17th century decided to improve the word by putting back the Latin 'b' to make SUBTLE, but as with other words treated like this [for example, doubt] the letter remained silent. In a way, that silent letter makes the word even more ... subtle.

··· *Suits* ···

SUITS

A SUIT is a derogatory way to describe the stereotypical Corporate Man, **who has nothing to define him but his working clothes** – a matching set of trousers, jacket, and possibly a waistcoat. It's similar to saying, "THESE GUYS ARE A BUNCH OF STUFFED SHIRTS." If you were to comment that, "THE BANK SENT A LOAD OF SUITS to our meeting", it suggests that those attendees **contributed nothing**, other than arrogance and the cut of their elegant, pinstriped Armani outfits.

Of course, not all suits are expensive and custom-made. When someone remarks, "SHE WAS ALL OVER HIM LIKE A CHEAP SUIT", they are referring to **a woman being publicly touchy-feely and possessive towards a man, when he does not necessarily desire that attention**. She is clinging to him like the fabric of a sleazy garment – in our imaginations, no doubt of a garish pattern and tasteless design. The earlier (1920s) expression about being "all over" someone – meant besotted by them, and usually making a public spectacle of this obsession, as in, "He was all over her like a rash". Recently, this "all over" phrase has increasingly been applied to celebrities who attract unwanted attention and adulation from the media or their 'groupies'.

There are two different ideas at work inside the word SUIT. The first concerns **dressing in a sort of 'uniform' that defines who you are or what group you belong to**; in other words, showing some sort of collective likeness and identity. The second concept is about **attracting admirers or followers**. The earliest source of both SUIT and SUITE is the Latin *secutus,* past participle of *sequere,* meaning to follow, and right away, it began to mutate into words that expressed that idea of attending, adhering to, or trooping along behind someone or something – as in *sequel.*

Now… imagine England after 1066, with all those elegant Norman lords showing off their power and status. They would go hunting (or, indeed, off to battle) with retinues of courtiers and servants dressed in similar colours and decorated with their Lord's emblems – very like a bright, noisy, living SUIT of playing cards – a term employed in gaming from about 1529. In their day the whole, jolly band would have been described by the Anglicised word *siwete,* from the Old French *siute* – so it's a tiny step from a SUIT to a SUITE or vice versa, to mean **a similarly dressed or liveried group of attendants, courtiers, or huntsmen**. Well SUITED for their task, away they'd gallop in PURSUIT of the quarry. There's that SUIT again!

And, by the way, if you've ever heard a Frenchman in a bit of a hurry order his lunch, the waiter will probably assure him of swift service by replying, "Tout de suite, monsieur." We translate that as, "At once, sir", but in times past it also meant, **"Everything will follow/arrive consecutively"**, so the poor customer might conceivably receive twelve separate plates, each containing a single oyster. Mort de rire!

In time, a SUIT also denoted **a feudal tenant's obligation to attend court** – usually to swear an oath of fealty or pay taxes on his pigs and hovel. Oh dear! It was never good news when you received a summons to go to Court, and it still isn't! Now you can begin to understand how the legal action of SUING or bringing a LAWSUIT was born. It's all about **pursuit through the courts of justice**, as if chasing down your prey with petitions and writs instead of crossbows. Even though these SUITS are only made of words, someone always gets speared and feels the pain.

A gent who PRESSES HIS SUIT with **sweet nothings and promises** rather than an iron and ironing board, is said to be a SUITOR; it's flattering for a lady to have more than one of these, even if she ultimately turns them all down. However, in business, being a SUITOR means a firm is making **outwardly friendly approaches to another company in hopes of buying it**.

That brings us neatly back to SUITS symbolising the world of commerce or management; this slang has only been part of our language since 1979, whereas the very useful **portable carrier** for SUITS and other garments – a SUITCASE – has been doing good service ever since 1902.

SWADDLE

The word SWADDLE was first used in England in the 14th century, but it comes from a much older word linked to SWATHE, that has very much the same meaning. The practice of wrapping an infant tightly in strips of cloth, about six inches wide, in linen or wool, goes back a very long way, and was known in Biblical times. Just think of the story in the Gospels of the baby Jesus at his birth, being SWADDLED and laid in a manger. Babies were commonly kept in these wrappings for days or even weeks on end without changing them. How uncomfortable – not to mention smelly.

The word means to envelope, to wrap, and can be used in a number of metaphorical constructs. You can certainly SWADDLE someone in **love and care**, but you could also be SWADDLED in **a sense of outrage**, or have **the sense of being constrained and held in** by being SWADDLED.

In fact, although the practice can be comforting for babies, from the 17th century the practice was challenged. Tight binding of a child's limbs, head and body is no longer considered acceptable practice, except for brief periods, as it can constrict the blood supply and distort the limbs or skull bones.

TOFF

Oxford University probably gave us TOFFS, a slang word used to describe rather superior people, often with a hint of arrogance in them. It comes from 'tuft', which was a local term for **a titled undergraduate**, because they wore gold tassels on their caps instead of plain yarn. It wasn't even all such undergraduates either; only the sons of those peers who had a vote in the House of Lords. Such prestigious young men became known as Tufts themselves, and even had a whole lot of admirers who became known as Tuftehunters.

In the 20th century, the general idea of snootiness was extended to TOFFEE-NOSED, and a more vulgar origin for the term was suggested, to wit: that it originally described **brown drippings from the nose of a snuff taker**. You see, only the richer undergraduates could afford snuff. This putative etymology is almost certainly an example of a later, manufactured explanation.

TOGS ☞ NAKEDNESS

We may refer jocularly to PUTTING ON OUR TOGS when we get dressed, and assume it is a modern word. Not a bit. Both the slangy expression ADAM AND EVE'S TOGS, referring to nudity, and 'togs' for clothes came from a 16th century word for a cloak, *togeman,* itself probably from the formal Roman *toga.* In modern urban slang, a TOG is not just clothes, but also three quite diverse things: **a small young man; the male genitalia;** or a short form of **'photog'** for **a paparazzi photograph.**

TOILET ☞ DRESS

Nowadays, we are likely to use the word TOILET almost exclusively to mean a receptacle, often made of porcelain, steel or wood, with a seat, a lid and plumbing, for purposes of defecation and urination. It may also refer to the room where this facility exists.

The earlier use described someone, usually by a woman, **sitting at a dressing table, making small adjustments to her attire and doing up her hair and face** until she had COMPLETED HER TOILET. The old term LA TOILETTE is still sometimes encountered, particularly in novels. In Old French, a *toile* or a *toilette* was a small cloth, usually made of a mixture of silk and linen for softness, originally used primarily as a wrapper for clothes, but later as a washcloth.

In the 17th century, the TOILETTE evolved into a cover for the dressing table. The word also meant a cloth draped over the shoulders while dressing the hair or wig – no doubt to protect a rich dress from the powder sprinkled on the hairpiece. The term still used for **the process and action of dressing up in your best**. In a rather macabre twist, LA TOILETTE also described the **final preparation of a prisoner for execution** – for example, Marie Antoinette's last toilette, assisted by her maid, Rosalie Lamorlière, before being taken to the guillotine. Later, the word would also be used for **cleaning up a street or a ship**, and is sometimes still heard referring to the cleansing of a hospital patient before an operation.

By the 19th century, the term TOILET had travelled to the USA and described **a room for dressing in, plus washing facilities** and it is this term, and the similar one 'lavatory', which is used commonly today. On occasion, Americans still refer to the 'powder room', also meaning the washroom or toilet, even though no one has worn powdered wigs there since the 18th century!

The term to make A TOILE is still used in dressmaking. This is **a mock-up of a garment**, created by a designer in a cheap fabric like canvas, cotton or muslin that can be manipulated and cut on the model, in order to try out the design before cutting into the expensive final fabric. There is also a printed textile used in interior soft furnishings known as 'toile de Jouy,' popular since the 1700s and usually showing romanticised scenes from that era. Incidentally, in the early 16th century, *toile* also describe a net stretched out to catch quail – yet another textile connection.

However, **toil** in the senses of hard work, and of stirring up strife, contention or confusion, had a separate aetiology to do with a Roman device for crushing olives, the *tudicula*.

TOWEL ☞ CURTAIN, MANTLE, SAIL and VEIL

A towel was originally a cloth to wet and wash with, not one meant for drying oneself. Our word comes from an old Germanic root, very possibly *thwakhlio*, meaning to wash. The Romans originally used the word *mantellus*, which not only denoted a towel, but also a curtain, tablecloth, or even a sail! That's stretching things a bit. Still later, *mantellus* came to mean an outer garment – hence mantle. The interesting "Oh, surely one bit of fabric should be able to do everything" approach seems apt for the pragmatic Romans. They always appeared to be dressed in household linen, making life in ancient Rome a useful theme for amateur theatricals, fancy-dress parties and the Grand Opening of new laundromats.

In India, the Sari is said to serve in exactly the same way. When in company, say, or on a train, it can be pulled over the lady's head for modesty or to provide seclusion while asleep, and is just as useful for wiping a child's nose or drying one's hands after washing.

You might use the phrase I NEED TO PUT A WET TOWEL ROUND MY HEAD to mean **you want to cool or clear your head in order to think**. The phrase is based on the idea that, otherwise, your brain might overheat – possibly from an excess of choler, one of the four 'humours' the ancients believed influenced our health and character.

TO THROW IN THE TOWEL derives from boxing, and means **to concede the match**. When the trainer chucks a boxer's towel onto the mat of the boxing ring, it's like waving a rather sweaty white flag of surrender.

Here's an old riddle, which creates a lovely, country feeling:

> "I washed my face in water
> That neither rained nor run,
> I dried my face on a towel
> That was neither woven nor spun.
> *(What am I?)*"
> Dew and sun.

TOWN AND GOWN

It's a tradition in Oxford and Cambridge to talk about TOWN and GOWN, a phrase distinguishing **the local residents – the town's folk**, from those **who live and work in the University, including the students** who wear black academic garb. The two groups are traditionally at each other's throats – sometimes literally.

TRAVESTY ☞ DRAG and VEST

Exchanging clothes with someone as a disguise is something aristocrats in past centuries did during times of conflict to avoid capture and heavy ransom payments for their freedom. The practice was known as *travestire* in Italian. This mutated into French *travestir*, meaning to ridicule, possibly because the toffs looked silly in their servants' gear.

This gives us our word TRAVESTY, meaning anything **absurd or outrageous and unacceptable**, such as a TRAVESTY OF JUSTICE for a trial that is felt to be **flawed and unjust**.

In English, A TRAVESTY is also **a burlesque** of a serious work of literature or art, or it can describe any bad imitation – again harking back to *travestir* meaning **to ridicule**.

TROUSERS ☞ BREECHES and PANTS

This interesting garment has a very long history. The Celts and Gaels, who often had to endure stiff Atlantic breezes and chilly winters, worked out a nifty way to keep warm. Our word for trousers came from the Irish *trius* and the Scots Gaelic *triubhas* which were close-fitting shorts. In the 16th century, these leg warmers crossed the border and entered the English language as trouse or trews in the singular – subsequently made plural as TROUSERS. They are still sometimes lightheartedly called TREWS, presumably from the older dialect word, now understood even by Sassenachs.

> "Diddle diddle dumpling, my son John
> Went to bed with his trousers on,
> One shoe off, and one shoe on,
> Diddle diddle dumpling, my son John."

"Diddle diddle dumpling" was the street cry of the hot dumpling sellers in the streets of London, who nabbed the phrase from a catchy 19th century music hall song. The rhyme is often sung before bedtime by a parent who sits with legs crossed in order to give a young child a 'horsey ride' on one rhythmically bobbing foot.

Here's a Bee Riddle from the early 18th century:

> There's a little short gentleman,
> That wears the yellow trews,
> A dirk below his doublet,
> For sticking of his foes.
> Yet in a singing posture,
> Where e'er you do him see,
> And if you offer violence
> He'll stab his dirk in thee.

This is, of course, another male chauvinist rhyme, since 'gentleman' bees and wasps don't sting. The bee's stinger evolved from an ovipositor and only females have those, but this fact was not generally known when the verse was composed.

Now, what about a song from the World War I that we Brits used to sing to keep our spirits up:

> "And the moon shines bright on Charlie Chaplin,
> His boots are a cracklin'
> For the want of blacknin'
> And his little baggy trousers need some mendin'
> Before we send him
> To the Dardanelles."

What about TROUSERING AN OBJECT? That means **to steal it**, by **sneaking it into your trouser pocket**.

WE CAN SEE WHO WEARS THE TROUSERS, or WEARS THE PANTS. It is possible that this use of a metaphor to suggest **a bossy wife and a browbeaten husband** may come from the early 17th century. It relies on the stereotypical idea of men always attired in trousers and women in skirts – with no recognition of the Scot's kilt or, indeed, a variety of trousers or trouser-like garments for women. This expression was said to have started as a 1606 proverb: "She that is master of her husband must weare the breeches."

Outspoken and capable wives are still said to WEAR THE TROUSERS when they take the man's traditional role as boss. In our time, it is hardly appropriate, since we have unisex clothing and have absorbed immigrant cultures in which women's trousers are a sign of their modesty. Back in 1933, Marlene Dietrich daringly sported a mannish tweed suit with turn-ups when she visited Hollywood, and Americans could already buy paper patterns for making wide-legged pyjama-style trousers like those worn by movie stars. In Britain, such things were really not considered completely acceptable till after WW II but the Land Girls and other women of WWI, who worked in traditionally male jobs, refused the inconvenience of skirts for such tasks. (See also BLOOMERS, DRAG, and PETTICOAT)

CAUGHT WITH YOUR TROUSERS DOWN! Pretty obviously, this means you are in a particularly vulnerable state! If you CATCH SOMEONE WITH HIS PANTS or TROUSERS DOWN, you've surprised him in **an embarrassing situation** or revealed him at **a disadvantage**.

In England, ALL MOUTH AND TROUSERS, was originally a North Country saying for a typical loud and gutsy type of **man, who stands no nonsense** and sticks to **a stereotyped male role**. He was said to be "all mouth and trousers" because he kept his mouth so far open that his torso couldn't be seen. Later on, the idea mutated to signify he was all bluster, and the saying is usually rendered as, ALL MOUTH AND NO TROUSERS, suggesting someone who can **talk big**, or **smarm and seduce in speech**, but well – to put it nicely, **can't deliver, especially not in the trouser department**.

TURBAN

Did you know that turbans are actually tulips? It isn't really surprising, because in a breeze, a group of these brightly-coloured spring flowers looks much like nodding, turbaned heads. A turban is made up from a long sheet of fabric, often silk, which can be wound around the head in a number of elaborate and colourful ways.

The word derives from the Persian *dulband,* which signifies both the wraparound head cloth and the flower that it resembled. This became *tuliband* in Turkey, and was taken into early modern French as *tulipan* and then *tulipa* in modern Latin, ditching the headgear meaning along the way and keeping only the botanical sense as in our word TULIP.

Meanwhile, the Turkish *tuliband* also morphed into *tulbend,* nipped over to Italy where the natives called it a *turbante,* and then sauntered into France, becoming *turban.* Finally, it crossed La Manche and settled down in Britain as TURBAN. All of which goes to show that both plants and words may develop in rich variety when they are transplanted to the fertile ground of other countries and cultures.

It is said that in 17th century Holland there was an economic 'bubble' caused by TULIP MANIA, known as *tulpenmanie,* when people started **a bidding war for exotic tulip bulbs**. In March 1637, at the peak of this wild speculation, some individual tulip bulbs sold for more than 10 times the annual income of a skilled artisan. As far as I am aware, no such frenzied enthusiasm ever arose for TURBANS, though the wearers have certainly always been very wrapped up in them.

TURNCOAT ☞ COAT

Someone who TURNS HIS COAT has **changed allegiance, betrayed his comrades and deserted his leader**. In other words, he has **gone over to the other side**.

There is a story that a Duke of Saxony, whose land lay on the border with France, hit on the idea of a reversible coat – white one side, blue the other, so that he could quickly appear to be on the winning side whenever necessary. As a result, soldiers of Saxony were sometimes derisively called 'Emmanuel Turncoat.'

I'm not sure about this story. It bears all the marks of a fable invented to insult a bunch of foreigners. What's my alternative? In folklore, if you are at risk of being abducted or bewitched by the fairies, one of the recognised ways of saving yourself is to take off your coat and turn it inside out. I wonder if the two things have something of the same origin? Both in a fairytale and in battle, the TURNCOAT makes himself a stranger, virtually a different person, thereby acquiring a kind of power and new choices he lacked before. Even so, there is almost always a price to pay.

There was also a superstition in the 18th century that, when losing at gambling in card games, the way to regain luck was TO TURN YOUR COAT AROUND and **wear it inside out**. Professional cardsharps took advantage of this and often turned their coats. They could conceal cards in false pockets or hidden slits. If their cheating was detected and they had to run away, they could quickly take on a different appearance and elude their pursuers by changing the colour of their outerwear via a speedy reversal of the coat. In that same period, spies are frequently described as having turned their coats to evade capture.

TWINSET AND PEARLS

The TWINSET is still around, though I haven't owned one for years. It was an admittedly rather pretty pair of knitted items consisting of a tight sweater, usually short sleeved, worn under a long sleeved cardigan in the same colour, usually left unbuttoned or buttoned only on the top button. Tasteful jewellery enlivened the neckline – traditionally pearls, or a discreet gold ornament on a chain.

To describe someone in a shorthand way as, "SHE'S VERY TWINSET AND PEARLS", sounds quite nice, but actually stereotypes a female as **a rather posh, tweedy countrywoman, the horse-riding, dog-breeding 'County' type, who is supposedly a right-wing snob**. The nuances of such prejudice and inverse snobbery in class-conscious Britain have a subtlety and nastiness that both puzzles and annoys more egalitarian visitors to our shores.

VAMPING A TUNE

VAMP was originally an Old French word *avantpie*, meaning 'before the foot' and referred to the part of the stocking that covered the upper front part of the foot. This meaning still applies to the relevant piece of leather used in shoemaking today. In time, the word became *vaumpe* in Anglo-Norman and by the 16th century, had segued into a verb meaning to replace the vamp on a stocking. Later on, it meant **to patch something up** to in a more general sense, and by the 18th century, **to extemporise**. So, when we describe a modern musician extemporising by VAMPING on the piano, are they really mending their socks?

In more innocent days, showing one's pretty ankles was apparently enough to seduce a man. And while it would be tempting to think of a 20th century flirt who VAMPS a chap by showing him her lovely, stocking-covered ankles, unfortunately this meaning has no textile connection. In fact, there's a far more sinister connection, for it derives from a Kazan Tatar word for witch, *ubyr* then *upyr*, which flapped its dark way from Russian into Hungarian as *vampir*. That's right – the bloodsucking, undead type of vampires popularised by the writer Bram Stoker in the 19th century. So gentlemen, beware. While you are checking out those gorgeous legs, this 'vamp' may be moving in for a quick love bite on your neck!

VEIL ☞ SUBTLE

The word VEIL comes from the Latin *velum*, which can mean a veil as we know it, a curtain, or even a sail; eventually, it became the Anglo-Norman *veile*. Indeed, the modern French term is *voile*, which we also use in English as the name of a fine, semi-transparent material used, among other things, for sheer curtains.

You can DRAW A VEIL OVER something figuratively, in order **to cover it up or ignore it.** When clouds or fog **hide the landscape**, it is often described as VEILED IN MIST. Of course, you can also REVEAL something, which basically means REMOVING THE VEIL from it.

TO TAKE THE VEIL means **to become a nun** in a Christian religious order, and comes from the clothing traditionally worn as a symbol of this change of status. In the Jewish religion, there is a precious cloth separating the body of the Temple from the sanctuary, which is seen as the earthly dwelling place of God; in ancient times, the High Priest was only allowed to pass beyond this heavy veil – supposedly about four inches thick and up to 60 feet high – once a year. According to the Gospel writer, it was this VEIL that was torn in two at the moment of Jesus' death, and for Christians, it signifies the accessibility to God for all people from that moment.

In another and decidedly worldly mood, a VEIL is a saucy addition to a hat used to half-conceal oneself, while possibly making 'come hither' looks with the eyes. It can be used as a full-face cover with formal dress when someone is in mourning, or to disguise the ravages of time by those who are vain about such things. For such a flimsy thing, a veil is really a very useful article of dress.

Not surprisingly it figures prominently in proverbs, particularly in the East where veiling is a fact of life for many women. A Moorish proverb goes, WHEN YOU SEE YOUR WIFE BLOWING HER NOSE WITH HER VEIL, DIVORCE HER, though I'm not quite clear why. Even more sombre, this comes from Saudi Arabia: A GIRL POSSESSES NOTHING BUT A VEIL AND A TOMB.

There is a much more pleasant Arabic saying which suggests that modesty does not reside in veiling the face of the woman, but in **the discretion of the observer** – THE EYE HAS A VEIL.

The deep significance of the veil is often revealed [UNVEILED] in legend and literature. The legend of Pyramus and Thisbe comes from the Classical author, Ovid, and was retold by Shakespeare through characters who act out the story in 'A Midsummer Night's Dream,' In the old Greek story, Pyramus, the hero, arranges an assignation with the heroine, Thisbe, but arrives to find only her bloodstained veil lying where a lion has mangled it. Believing her dead, Pyramus kills himself. No doubt, Shakespeare also recalled this tale when writing about the tragic reunion of those 'star-crossed lovers,' Romeo and Juliet.

VEST ☞ INVEST YOUR MONEY and TRAVESTY

Charles II introduced the fashion for wearing what we British would call a waistcoat, although in the 17th century, it was a much longer affair, fitting right over the hips. In the BT days (Before Television) the Yanks – World War II servicemen passing through our country – introduced many of us to US lingo. I can still remember feeling a slight sense of shock and disapproval at learning that Americans sometimes walked around in public in 'vest and pants', as I imagined the whole nation to be ambling about in their undergarments. Of course, a 'vest' in America is a waistcoat, not a singlet or undershirt, and 'pants' are trousers, not knickers. Surprisingly, this confusing Transatlatic divergence only happened in the 19th century; that's when we in the UK started confining the use of these words to intimate underclothing.

The word VEST, however, may go back millennia to an Indo-European root – either *wes* or *wazjan* meaning to wear or be dressed. This evolved into the Greek *heanon* and then *Hestia* (of whom more in a moment), and was taken up in Latin as *vestire* – to clothe. Later, this spawned both the French and Italian *veste*, from which our English word VEST emerged.

Originally, British folk used VEST to denote **any robe or gown**, and not simply the waistcoat – hence VESTMENT and VESTRY, words nowadays associated mainly with the clergy. The VESTRY is the small **room in a church** where the priest or parson changes into his VESTMENTS or **religious garments**. The VESTIBULE, meaning a **small antechamber** or **lobby** near the main entrance, also comes from this root. It can be used in a secular setting, but is more often in the church again.

Another term incorporating VEST is the term TRANSVESTITE for someone who cross-dresses by **wearing clothing of the opposite sex**. It only came into common use during the early 20th century in Germany. Think of the show 'Cabaret' and all the exotic, erotic goings-on in inter-war Berlin.

Now a bit about Hestia. She was the Greek goddess of the temple's sacred sacrificial flame, kept alive by **the priestesses** known as the VESTAL VIRGINS – there's that word VEST again. She also presided over the hearth, centre of the home and the source of both heat and nourishment. It's because of her that a famous British brand of matches was called **Vestas**, and there were some lovely little silver and gold matchboxes produced during Edwardian and Victorian times called **Vesta cases** that are eagerly collected today. See how far that versatile VEST can expand?

Recently, researchers in Cardiff, Wales, showed that getting chilled really does reduce your immunity so that you may catch more colds. There you are. Your mum was definitely right when she insisted you wear a nice warm VEST next to your skin on a bitter day – and you didn't listen to her when you were growing up, did you?

WALTZING MATILDA ☞ BLANKET

Blankets keep us warm. But another use for a blanket is to wrap up and carry belongings or "swag", as Australian travelers known as swagmen once did. These itinerant workers, perhaps only one degree up from tramps, would roll all their worldly goods inside a blue or grey blanket and tie the loose ends with rope or cord, so the whole thing resembled a giant Christmas cracker or bonbon. Then they would hoist it over their shoulders like a knapsack as they went around the country looking for seasonal work.

The term WALTZING MATILDA originally may have referred to this blanket roll. When the swagman hit the road, he went 'on the waltz,' a phrase from German slang meaning **to take to the road**. The MATILDA was also the name for an **army greatcoat**, either carried rolled over the shoulder or used like the blanket to envelop belongings.

Whatever the exact details, the singer songwriter, A. B. (Banjo) Paterson was inspired to pen the famous song WALTZING MATILDA while out in the Queensland countryside in 1895. The story of the swagman's theft of a sheep and his defiant end, choosing to drown himself in a billabong or pool rather than be captured by the authorities, may have been based on a local incident. Over time, this song has become a kind of unofficial Australian National anthem, and expresses both the Aussie's independent mindset and love of freedom.

The term swag now tends to mean 'stolen goods' and, in England, was used in that sense from the early 16th century. It was also a verb that suggested causing something to sway or sag, derived from a word of Scandinavian origin for a bulging bag. Then the word was transported – probably along with a shipload of prisoners – to Australia, where, along with terms like bundlemen and swaggers, it came to mean the goods carried by folk who lived in the wilds.

WELSH HATS

There is a splendid story about the last time Britain was invaded, which was not, as most people think, in 1066 by William the Conqueror, but in 1797, when soldiers from the Revolutionary French Army landed at Fishguard in Pembrokeshire, West Wales. This small troop of invaders took over a few farms and shot a grandfather clock by mistake. Then they discovered a lot of wine 'liberated' from a Portuguese ship wrecked the previous week, and got very, very drunk. One feisty woman, Jemima Nicholas, afterwards known as Jemima Fawr (her byname meant 'the great' in Welsh) rounded up a dozen of them with a pitchfork, marched them to St Mary's Church, locked them in, and went back for more.

Next morning, it is said that the French looked up at the cliffs and saw what they believed to be masses of Redcoats marching up and down. In fact, these "Grenadiers" were the local Welsh women in their cloaks of scarlet flannel and wearing tall black hats, marching round and round to foster the idea of many troops. The French soldiers, bedraggled and badly hung over, assembled on the beach at Goodwick Sands with their American-Irish commander, Colonel Tate, and surrendered, believing themselves to be hopelessly outnumbered by British soldiers.

Although the truth may have been embroidered not a little, it is a satisfying idea that women's fashion, so often seen as trivial, could fool a body of professional fighters and thwart their aggression!

This story of Jemima Nicholas and the Last Invasion of Britain was celebrated in a 100-foot embroidered panel, created by volunteers working for two years and exhibited in Fishguard to record the bicentenary of the invasion in 1997. Today it has a fine home in a gallery at Goodwick Sands. If you're passing, why not pop in for a look?

··· *Welsh hats* ···

WIGS ☞ HAT and RUG

Do wigs count as headgear, or more as prostheses? They are, of course, the artificial mini-rugs of hair or something similar which can cover the sensitive baldness of the pate, usually of a man. Of course, they may also be worn by anybody for a theatrical performance or when dressing up for a party or GALA! [qv].

By 1680, Judges and Barristers of the British Court system wore WIGS made of horsehair, with little curls on the sides and tails down the back, and they still do today. Courtrooms in those days were so filthy that lice were everywhere, so horsehair was selected to combat the problem, and most legal gents wore them over shaved heads – just in case. Being innately conservative, judges thought the new wigs looked "coxcombical" – or a bit too flashy – and forbade young advocates to plead while wearing them. But eventually, the old boys unbent a bit and everyone adopted them.

Originally, the term for wig was PERIWIG from the French *perruque*. Where this word arose is unknown, but possibly from *perroquet* or parrot. What a marvellous picture of the legal system that gives us!

The shortened word WIG came in during the 17th century, when a term like DASH MY WIG! was a mild form of swearing. A WIGGING was and is still a sound **scolding or rebuke**, such as one might receive from a stern magistrate or judge.

The bigger the wig, the more important the person. This rule of thumb (or hair) comes from the days when wigs were expensive both to buy and to maintain, thanks to a government tax on powder. Those who flaunted their wealth and status by wearing long or "full-bottomed" wigs naturally attracted their share of enmity from humbler sorts who could not afford such luxuries.

British judges and some officials in the House of Lords still wear such things, at least for ceremonial occasions, and that is why we tend to call **important people** BIGWIGS. This must be said, of course, in the gently derisory tone that the Brits have perfected to cut their betters down to size. If the Important Person in question gets upset, they may be told, "KEEP YOUR WIG ON" which, as with KEEP YOUR HAT ON, means '**calm down!**'

An alternative modern term for the WIG is A RUG [qv] which is **a hairpiece** worn by follically-challenged men and derided for its resemblance to a cheap chunk of shag carpet.

In the 1950's, the terms TO FLIP YOUR WIG or TO WIG OUT became part of young people's slang to mean **wildly excited and behaving hysterically**, either through high spirits or intoxication.

WOODEN OVERCOAT

A coffin – 'nuff said.

ZONE ☞ BLOUSE

A ZONE is a word for a region, as you no doubt know – such as the various climate ZONES encircling the earth. However, it can also be applied to **any region with definite limits**, such as the **facets of a crystal**. The word came into regular use during the late 18th century, but had been used metaphorically since at least 1500.

In the ancient classical world, only two words have been reconstructed for clothing, one being the ZONE or zostra for a belt. The other is a generic word Homer uses for the garment worn by his wife and Queen of deities, Hera, when she visits Zeus. This is the *heanon* that we have already met – with regard to BLOUSE and as the root word for a vest, probably once worn like a blouse, rather than as an undergarment or waistcoat.

Back in 9th century BC Greece, the ZONE or *zostra* meant a broad, flat belt, worn low on the hips by young virgins – but only on their wedding day. It was removed by the husband in a formal ceremony after the marriage vows were made. The everyday belt that girls and women wore was called a *cingulum*, which was cinched just beneath the breasts. Art historians and archaeologists still talk about the zone or belt on carvings of Greek goddesses such as Aphrodite. Well, Aphrodite may have lost her nightie, but at least she's kept that belt!

The ZONA IGNEA or 'Fiery Zone' was a medieval term for the viral skin condition we know as **shingles**.

In American football the END ZONE is the **bit of ground at the end of the pitch where a touchdown is scored**. So, this END ZONE perhaps marks a good point to say goodbye to our non-exhaustive (though fairly exhausting) alphabet of Words with a Textile Origin and perhaps we can all feel that we are winners.

Thank you, my friends, for exploring all the way to

THE END

... Index ...

This is an index of the words and phrases discussed in the book, plus the various names mentioned and the topics covered. **Bold** page numbers indicate main entries. Some phrases have been indexed using only their key words, and most of the phrases using that word will be found in its main entry. Source words in languages other than English have not normally been indexed.

abandannad 6
Abram 131
accent, Irish 24
Achilles 56
Adam & Eve 14, 37, 102, 143
address 57
admiration 82
Aesop's fables 40
airing dirty laundry in public 92
alcohol
    blue ribbon army 17
    bracer 21
    nightcap 103
Alexander the Great 134
Alice in Wonderland 84
all dressed up and nowhere to go 58
all hat and no cattle 83
all mouth and trousers 83, 147
all over someone 141
American Indian myths and legends 15
anchor, sheet 124
anger 44, 84, 119, 124
animals, poaching 111
anorak 1
Antoinette, Marie 144
ants in your pants 1
Aphrodite 156
Apollo 56
appall 104
applause 121
apple pie bed 1
apple pie order **1–2**
apron **2**, 11–12, 22
architecture, filet 63
armour **2–3**, 93

army/soldiers
    blanket drill 15
    blue ribbon 17
    carpet knight 32
    colours 46–7
    dressing troops 57
    dud shells 59
    flak jacket 89
    friendly fire 45
    leatherneck 93–4
    mufti 101
    rally round the flag 64
    retreating 5
    top brass 85
    undertunic 26
    women soldiers 56–7
arrest 44, 74
asbestos **3–4**
ashes, and sackcloth 123
at the drop of a hat 85
Austen, Jane 29–30
Australia 13, 16, 21, 66, 101, 106, 107, 122, 132, 139, 152–3

bad hat 83
bag **4–5**
    bag lady **5**
    baggage 5
    bags I! 4
    carpet bag 33
    Christmas stocking 35
    handbag **79**
    in the 4
    left holding the 4
    letting cat out of 4
    manbag 5

poke 109
portmanteau **113**
satchel 122
swag 152, 153
baize 26
Baker, Robert 108–9
bamboozle **5–6**
bandanna **6**
bandbox **6–8**
banner **8**
baseball mitt 99
bast **8**
bastard **8–9**
baste **8**
bathing costumes 5, 14
bathrobe 118
batt/batting **8**
beard, capuchin's 31
bed **9–10**
    apple pie 1
    curtain lecture 49
    duvet 47, 50, **60**
    feather 10
    make your bed and lie on it 10
    quilt 47, 49–50
    sheets 124
    snug as a bug in a rug 121
    tucked up in 13
bedfellows 10
bedraggled 55
bee
    riddle 146
    in your bonnet 18
bells
    and cap 29
    horse trappings 30

belt **10–11**
  belt up! 11
  black 11
  and braces 11
  cinch 35
  embroidered 10
  getting it under your 11
  girdle 71
  hitting below 11
  tightening 11
  zone 156
belter 11
Bennett, Alan 136
Beowulf 76
bib and tucker **11–13**
Biblical references/stories 14, 40, 56, 97, 102, 131, 142, 143
bibs 12
bicycle 68
big girl's blouse 17
bigwigs 155
bikini **14**
birthday suit 14, 101
black belt 11
black monks 46
blackshirt 45
blanket **14–16**, 152
  born on wrong side of 9, 15
blood, on the carpet 33
bloomer, to make a 16
bloomers **16**
blouse **16–17**, 156
blow 41, 48, 50, 79, 83, 93, 135
blowzy 17
blue collar worker 44
Blue Peter 66
blue ribbon 17
blue sky **17–18**
blue-on-blue 45
bluestocking 45, 138
Bobby soxers 144
bodice 18, 21, 95
bodice-ripper **18**
body
  does my bum look big in this? **54–5**
  nakedness 14, 61, **101–2**, 138, 143
  protective covering 2, 11, 82, 89
Bogey Man/Bogart 81
bolster 110
bombing, carpet bombing 33
bonnet **18–19**

books
  bodice-ripper **18**
  dust jacket 89
  fake reviews 135
  pillow book 110
  pocket edition 112
  portmanteau 113
boot **19–20**, 127–8, 129
booting out 19–20
bootless 20, 134
bootstraps, pull yourself up by 20
booty 19
bottom
  does my bum look big in this? **54–5**
  eat my shorts 61
bowler hat 85
boxing 11, 114, 145
boys, to be breeched 23–4
Boz'll 23
bracelet 21
bracer 21
braces 11, **21**
brainstorming 65
brass hat/cap 85
brassiere/bra **21–2**
brat **22**
breeches **22–4**
breeks 22–3
bribe 76
Britain
  Fascist Party 45
  last invasion 153
britches **22–4**
brogue **24**
buckle **24–5**
budget **25**, 113, 130
bully 119
bureaucracy **25–6**, 85
burgeoning 26
burial
  coffin 156
  pall 104
  shroud 124, **131–2**
  white gloves 75
  winding sheet 124, 131
  in woollen shroud 14–15
  see also death
burning
  blanket to get rid of fleas 15
  your bra 22
Burns, Robert 51
bursar/bursary 113–14

bus, flagging down 65
busk 95, 96
buskin, brother of the 20

Cailleach 104
camiknickers **26–7**
camisole 27
camouflage 36
camping, canvas tent 28
canapé 27–8
canopy **27–8**
canvas 27, **28**
cap **28–30**, **103**, 143
caparison 30
cape 30, **30–1**, 85, 94
cappella 34
cappuccino **31**
Capuchin monks 31, 76
car
  bonnet 18
  coat 90
  hood 86
  limousine **94**
  red flag laws 65
  shoes hung on 128
cards see playing cards
Carlyle, Thomas 74
carpe diem 32
carpet **32–4**
carpetbagger 33
Carroll, Lewis 84
Cartwright, Nancy 60
Casanova 111
cassock 31
cast not a clout till May be out 41
cat
  cat's pajamas 115
  letting out of bag 4, 109–10
  Rutterkin 75
Cato 130
Celtic myths and legends 5, 83, 104
chain mail 2, 73
Chancellor of the Exchequer 113
chapel **34**
chaperone **34**
chaplain 34
chaplet 34
Charlemagne, King 3
Charles II, King 14, 151
Charles, Prince of Wales 89
cheating 13, 18, 40, 42, 86, 102, 149

158

children
  bibs 12
  brat 22
  breeches 23–4
  coloured clothing 46
  dressing up 58
  illegitimate 8–9, 15
  protecting babies 133
  rug rat 122
  security blanket 16
  swaddled 142
children's rhymes 10, 22, 25, 50, 87, 98, 103–4, 106, 107–8, 111, 112, 128, 139, 146
  see also popular songs & rhymes
chlamydia 36
Christmas
  carol 43
  stocking **35**, 138
  tree, dressed up like 59
cinch **35**
Cinderella 72, 129
Cistercian monks 46
civil service 33, 85
cleaner/charlady 99
clergy
  chaplain 34
  clothing 31, 67, 151
  man of the cloth 39
  unfrocking 67
clever clogs 37
cloak **36**
  brat 22
  caparison 30
  cape 30
  cappuccino 31
  and dagger 36
  domino 55
  escape 61
  mantle 96–7
  palliative 104
  portmanteau 113
  sunrise/sunset 15
  togs 143
cloaking device 36
clogs **37**, 122
clootie dumpling 41
cloth
  cloth ears 39
  coloured 45
  cut according to your cloth 36
  frieze 66–7
  man of the 39

motley 99
pane 50
ribbon 116
subtle 139
toile/toilette 143
voile 150
clothes horse 39
clothing **37–40**
  asbestos sleeves 3
  blouse **16–17**
  cast not a clout till May be out 41
  clothes make the man **40–1**
  colour **45–6**
  costume 79
  cuff **48–9**
  cut 116
  disguise 53–4
  does my bum look big in this? **54–5**
  dress **57–9**, 97
  dress-down 58
  duds **59**
  dungarees 12
  eat my shorts **60–1**
  escape 61
  fits like a glove 76
  frippery 67
  gala 68–9
  given to the Brownies 39
  habit 78
  if the cap fits, wear it 30
  inside out 54
  laundry 92
  like a Nun's nightie 103
  men wearing women's 55–7, 89, 107
  pinafore dress 12
  pocket 109, 111, **111–12**
  ribbons 116
  robe 19, **117–18**
  ruffs 6, 118–21
  sack/sacque 123
  sackcloth and ashes 123
  shirt 51, 82, **124–7**
  Sumptuary Laws 99, 118–19, 120
  to be clothed with 39
Clotho, Greek Goddess 39
clotted cream 41
cloud/mist 86, 150
clouter 41
clouts **41**
clumsiness 101

coat **42–3**, 89–90
  of arms 2
  army greatcoat 152
  cut according to your cloth 36, 43
  frock coat 67, 68
  fur coat, no knickers 68
  with hood 85
  of mail 2
  turncoat **148–9**
  white 46
coattails, climbing/riding on 42
coatum 90
cobblers 128, 130
coccyx 50
coffee, cappuccino 31
coffin 156
collar **43–4**
  band box 6
  pickadils 108–9
  ruff 118–21
colour
  clothing **45–6**
  motley fabric 99–100
  show your true 47
colours (flags) **46–7**, 64
combinations 106
Commedia dell'Arte 105, 119
computers, rebooting 19
considering cap 29
cope 34
corset 16, 18, 21, 24, 90, 95–6
costume 79
cotton socks 135
counterpane 47, 50
counterpoint **47**
country, pride in 64
court
  court cards **48**
  lawsuit 142
  wigs 155
Courtneidge, Cecily 82
criminals 86, 130
Crockett, Davy 21, 23
cross-dressing 55–7, 89, 152
cross-gartering 71
crying 12
cuff **48–9**
cunnilingus 100
curtain **49**, 144
cushion **49–50**, 110
cut a rug 33
Cutty Sark 51
cylinder jacket 89

dagger, and cloak 36
dancing
    cut a rug 33
    dandy 53
    line dancing 94
    ring dance 137
    ring-a-ring-o'roses 112
dandy **51–3**
danger, flagging up 66
dash my wig 155
deadline 94
dealer's shoe 131
death
    curtains 49
    hang up your hat 83
    hose 86–7
    sackcloth and ashes 123
    to pop your clogs 37
    to rob us of our loved ones 117
    veil 150
    waiting for dead men's shoes 129
    white gloves 75
    wooden overcoat **156**
    see also burial
deception 13, 18, 40, 86, 102, 149
Dee, Nannie 51
deep pockets 112
desk, writing 25–6
desolation 20
detail 116
Devil 47, 54, 119, 128
diamond geezer 54
dibs 4
Dickens, Charles 83
Dietrich, Marlene 147
dinner jacket 90
disguise 36, 40, **53–4**, 85–6, 98, 145
does my bum look big in this? **54–5**
dog
    dressed up like dog's dinner 59
    handbag dog 79
    pillow dog 110
Dominican monks 46
domino **55**
don't get your knickers in a twist 92
down at heel 130
drag **55–7**

drawers 91
dress **57–9**
    mantua 97
    robe 117
dress coat 90
dress-down 58
dressage 57
dressed to kill 57
dressed up to the nines 57–8
dresser 57
dressing down 33, 58
dressing up 58, 59
drop a line 94
duds **59**
duel, challenge 73, 79
dungarees 12
dust jacket 89
duvet 47, 50
    duvet day **60**
Dyfed, Lord of 75

ears
    cloth ears 39
    pig's 114–15
Easter bonnet 18
eat on the cuff 48
eat my hat 83
eat my shorts **60–1**
education
    bursar/bursary 113–14
    flying colours 64
    town and gown **145**
    university diploma 123
Edward III, King 70, 72
eggs, poaching 111
Elijah, passes mantle to Elisha 97
Elizabeth I, Queen 81, 99, 120, 137
embroidery
    belt 10
    cuff 48
    frieze 66–7
    smocking 68, 134
emotions
    and colour 46
    masking 36, 98
    showing 123
    wearing your heart on your sleeve 133
emperor's new clothes 102
end of the line 94
end zone 156
English myths and legends 19, 23, 107
escapade 61

escape **61–3**
escapement 61
Eve & Adam 14, 37, 102, 143
exquisites 51
eyes, sheep's 123

fables
    Aesop 40
    pig in a poke 109
    Truth & Falsehood 102
fabric see cloth
fairs
    pig in a poke 109
    Royal Charter 74
fairy/folk tales
    Aesop's fables 40
    Andersen 102
    Beowulf 76
    Cinderella 72, 129
    clothing 39–40
    cross-dressing 56–7
    English 19, 23, 76, 81
    Grimm brothers 10
    Irish 71, 81
    shoes/boots 128
    turncoat 148
    Welsh 75
family, his hat covers his 83
fascia 9
fascinators 6
Fascist Party 9, 45
fashion
    bandbox 6
    bib and tucker 11
    bloomerism 16
    clothes horse 39
    dandy 51–3
    does my bum look big in this? **54–5**
    dressed up like a dog's dinner 59
    mutton dressed as lamb 59
    old hat 82
    sleeves 133, 134
fasteners
    apron strings 2, 22
    buckles 24, 25
    tie/strap 63, 116, 130
Father Christmas 35
fear 104, 129
featherbed, to 10
feathers
    in duvet 60
    in hat 53

160

feminists 22
Fenrir 117, 130
fertility 61, 69, 128
fight
    challenge to 73, 79
    get your shirt out 124
    my hat is in the ring 85
    truce 74
files, bastard 9
filet/fillet **63**
finances
    budget **25**, 113, 130
    cushion 50
    cut coat according to your cloth 36, 43
    put a cap on 28
    tightening your belt 11
    see also money
first impressions 37, 57
Fisher, Kitty 111–12
Fishguard 153
Fitch, Thomas 53
fits like a glove 76
Fitzroy 9
flag **63–6**
    all flags flying 64–5
    banners 8
    blue pennant 45
    Blue Peter 66
    colours **46–7**
    of convenience 65
    flag down 65
    flying false 5, 47
    nail your colours to the mast 47
    pirate 47
    sending messages 64, 66
    upside down 65
flag days 66
flagging down 65
flagging up 65, 66
flagship 64
flak jacket 89
flatlining 94
flax 61, 63, 94, 119
fleas
    pillow dog 110
    why burn blanket to get rid of 15
flip your wig 155
flounce **66**
Flower, Joan 75
flowers
    beds 10

carpet of 32
crown of roses 34
flag 63–4
foxglove 76
fly the flag 64
flying by the seat of your pants **66**
flying colours 64
folk songs see popular songs & rhymes
folk tales see fairy/folk tales
food
    capuchin's beard 31
    clootie dumpling 41
    clotted cream 41
    dressed 57
    eat on the cuff 48
    filet/fillet 63
    girdle/griddle 72
    to tuck in 13
fool's cap 29
foot see boot; shoe; sock
fops 51
foxglove 76
Franciscan monks 31, 46
freebooter 19
friendship 73, 129
frieze **66–7**
frill 66
frippery **67**
frock **67–8**
frock coat 67, 68, 90
fur
    fur coat, no knickers 68
    glass slipper 73
    muff 100
furbelow 66
furniture, dresser 57

gala **68–9**, 155
gambling 126, 130, 131, 149
games
    beat the pants off 105
    domino **55**
    playing cards 48, 131, 141, 149
gangs 85, 86
garment makes the man 40
Garrick, David 121
garter **69–71**, 91, 138
Garter, Order of 17, 70
gauntlet, to throw down 73
Gaye, Marvin 84
geezers 54

genitals
    muff 100
    naked 61
    ribbon 116
    smock 134–5
    tog 143
George and the dragon 72
get out, bag and baggage 5
girding 71–2
girdle **71–2**
give way 25, 30, 65
Given, Thomas 22–3
glass slipper **72–3**
Gleipnir 117
gloves **73–6**, 98, 99
goat
    hair shirt 126
    kid gloves 74
    scapegoat 63
gods & goddesses
    Aphrodite 156
    Clotho 39
    Hera 16, 156
    Hestia 16, 151, 152
    Norse 117, 130
going commando 102
going hell for leather 93
Goldsmith, Oliver 129
good luck 128
goody two-shoes 129
gray eminence **76**
Gray, Mr 45
gray/grey monks 46, 76
Greek myths and legends 16, 39, 55–6, 127, 129, 151, 152, 156
Greene, Ann 69
Greene, Mr 45
Greenland 1
Grendel 76
Greyfriars 46, 76
guidelines 94
Guilds of Merchants 78, 87, 93
guisers 54
guts, have your guts for garters 69

haberdashery **76–8**, 108
habit **78–9**
hair
    completing your toilette 143, 144
    curly 24
    filet 63

hairpiece 155
mantilla 97
mop 99
shirt 126
wigs 144, **155**
Hamlet 40
handbag 79
handcuffs 21, 49
handkerchief **6**, 41, 75, **79–82**
hands
    asbestos 3
    hand in glove 73
    iron hand in velvet glove 74
    muff 100
    on one hand or the other 75
    shake your mitt 98
hang up your hat 83
hang you out to dry 92
hanky-panky 80
Hapgood Burt, Benjamin 58
harness 30
harvest 32
hat trick 84
hats **82–5**
    bonnets **18–19**
    caps 28, 29–30
    chaperon 34
    cocked 83
    feathers 53
    hat box 6
    knotted handkerchief 81–2
    nightcap **103**
    turban **148**
    veil 150
    Welsh **153–4**
have your guts for garters 69
Hawthorne, Nathaniel 68
headland 30
headlines 94
healing 81, 131, 138
heart
    wearing your heart on your sleeve 133
    in your boots 20, 129
hemp 27, 28
Henry VIII, king of England 137
Hera, Greek goddess 16, 156
heraldic insignia 2–3
Hercules 56, 127
Hestia, goddess 16, 151, 152
hide (leather) 23, 92–4, 123
high hat 82
Hindi 6, 12, 61, 115

hit 41, 48, 79, 83, 93, 135
hoax 5
home is where you hang your hat 84
Homer 156
hood 31, 34, 55, 82, **85–6**
hoodies 86
hoodlum 86
hoodwink 86
hooliganism 76, 86
horses
    blue riband 17
    clothes horse 39
    dressage 57
    leather 93
    put your shirt on 126
    rich purse 114
    trappings 30
    see also saddles
hose **86–7**, 137
housecoat 42
Hulbert, Jack 82
Huxley, Aldous 76
Hyland, Bryan 14

India 15, 61, 68, 100, 102, 115, 120, 128, 145
insignia, heraldic 2–3
instinctive actions 66
intensifying words 9, 93
internet, fake identities 135
Inuit 1, 72
invest/investiture **87–9**
invisibility, cloak of 36
iron hand in velvet glove 74
Irving, Washington 92
Italy, Fascist Party 45
ITMA (It's That Man Again) 99

jacket 1, 42, 46, **89–90**
Jenks Bloomer, Mrs Amelia 16
jester 29, 99
jewellery 59, 149
Jolly Roger 47
jumper 42, **90–1**

Keeler, Christine 110
keep it under your hat 82–3
keep your hat/wig on 155
kerchief 80
kid gloves 74
kings, illegitimate children 9
kiss my chuddies 61
knickerbockers 92

knickers
    camiknickers 26–7
    don't get your knickers in a twist 92
    fur coat, no knickers 68
    knickers to you too! **91–2**
    red hat/shoes and no knickers 68
knight
    of the carpet 32
    George and the dragon 72
    Order of the Garter 70
knitting
    machines 137
    stockings 136–7
    twinset 149
knock into a cocked hat 83
knot
    in handkerchief 80, 81–2
    rug 32

lamb, mutton dressed as 59
laugh up his sleeve 133
laundry **92**
lavatory 144
lawsuit 142
leadership 64
leather 23, 63, **92–4**
leatherneck 93–4
lecture, curtain 49
legends see fairy/folk tales; myths and legends
Leoncavallo, Ruggero 100
liar, liar, pants on fire 106
lick your boots 20
lies/lying 5, 20
lifeline 94
lightning 124
limousine **94**
linen
    chemise 51
    handkerchief 81, 82
    household 2
    pall 104
    sheets 1, 124
liners, ocean 17, 113
lining **94**
Linus 16
lips, pucker/purse 113
liquorice shoelaces 130
list, laundry **92**
Little, Arthur D. 115
Livery Companies 78, 87, 93
Locket, Lucy 111

loins, girding 72
London
    clothing 85, 87, 120
    Georgian dandy 51, 53
    Guilds 78, 87, 93
    Petticoat Lane 107
    Piccadilly **108–9**
    theatre 65, 82
loose jumps 90, 95
loose women **95–6**
lounge coat 89–90

Mabinogion 75
macaroni 53
mad as a hatter 84
Madonna (singer) 63
magic
    carpet 34
    handkerchief 80, 81
    stockings 138
mainline 94
manbag 5
mankini 5, 14
mantelpiece/mantelshelf 97
mantilla 97
mantle **96–7**, 113, 144
marriage
    cross-dressing story 56
    garters 69
    hang up your hat 83
    she set her cap at him 29
    shoes on wedding car 128
    stockings 138
    tied to wife's apron strings 2
    who wears the trousers 147
    zone/belt 156
masculinity 2, 17
mask **98**
masque 98
masquerade 98
matches, Vesta 152
Matilda, Waltzing 152
memory 80
mending 8
menstruation 41, 65, 70
mental illness 46, 139
messages, sent using flags 64, 66
Miller, William 104
misers' mitts 98
mistake, making 10, 16
mittens **98–9**
Molière 126
money
    glove money 76

investing **87–9**
living on a shoestring 130
    pocket 112
    purse 113–15
    shirt on/off your back 126
    well-heeled 130
    see also finances
monkey, capuchin 31
monks
    Capuchin 31, 76
    Cistercian 46
    clothing 46, 67, 78
    Dominican 46
    Franciscan 31, 46
    gray/grey 46, 76
Montagu, Lady Elizabeth 45
mooning 61
mop **99**
morning coat 90
Moroccan proverb 123
Moseley, Oswald 45
mosquito net 27
motley **99–100**
muff **100–1**
muffler 100
mufti **101**
mummers 54
music
    counterpoint 47
    sheet 124
    vamping a tune 149
music hall 84, 91, 146
Mussolini, Benito 45
mutton dressed as lamb 59
myths and legends
    American Indian 15
    Australia 101
    Celtic 5, 83, 104
    English 19, 23, 107
    Greek 16, 39, 55–6, 127, 129, 151, 152, 156
    King's shirt 127
    Norse 117, 130
    urban legends 3, 44, 122
    Viking 23, 134

nail your colours to the mast 47
nakedness 14, 61, **101–2**, 138, 143
nap your bib 12
napery 2
Napoleon 74, 85, 133
NATO (North Atlantic Treaty Organization) 45

navy
    all flags flying 65
    colours 46
    flags 64
    flagship 64
    flying false flags 5–6
    lower the flag 65
    pocket battleship 112
necklace 8, 44
Nelson, Admiral Lord 64
newspaper 124
Nicholas, Jemima (Fawr) 153
night out 126
night-night 103
nightcap **103**
nightdress/nightgown **103–4**
Norse myths and legends 117, 130
novels, bodice-ripper 18
nuns, take the veil 150
nursery rhymes see children's rhymes

ocean liners 17, 113
Odysseus 56
old hat 82
old man 54
Oliver, Sweet Polly 56–7
O'Malley, Grace 81
Omphale, Queen 56
oompah, oompah, stick it up your jumper 91
Orczy, Baroness 107
Order of the Garter 17, 70
O'Shanter, Tam 51
Othello 80–1, 114
outskirts 132
overalls 12
overcoat 42, 90, 156
Ovid 151

padding 8
paint 42, 43
pall 104, 124
palliative **104**
paltry **104–5**
Pantalone 105
pantaloons 91, 105
pantomime, drag queen 55
pants **105–6**
    ants in 1
    see also trousers
paper
    foolscap 29
    sheet 124

163

parchment 9, 47, 92, 123
Parliament 4
pass round the hat 84
Paterson, A.B. (Banjo) 152
pawnbrokers 35, 37
Peanuts 16
pearls, and twinset **149**
Père Joseph 76
periwig 155
Persia 27, 31, 42, 115, 134, 148
petticoat 42, **106–8**
photograph, tog 143
Phrygia 66
Piccadilly **108–9**
pig
　in a poke **109–10**
　to make a silk purse out of a sow's ear 114–15
pigeon 31
pillow **110–11**
pinafore
　apron 2
　discipline 107
　dress 12
pirate 19, 47, 81
plants 10, 26
　see also flowers
playing cards 48, 131, 141, 149
pledge, glove 73
Pliny the Elder 4
poaching 111
pocket 109, 111, **111–12**, 131
poetry & verse quotations 20, 22–3, 51, 54, 70–1, 74, 95–6, 117, 129, 132, 137
pointless exercise 134
poke 109, 111, 113
police
　to collar 44
　to cuff 49
politics
　canvassing votes 28
　carpetbagging 33
　Parliament 4
　petticoat government 107
　public purse 113
　throw hat into the ring 85
　to handbag 79
Pompadour, Madame de 107
pop your clogs 37
popular songs & rhymes 2, 14, 21, 25, 40, 43, 45, 56–7, 69, 84, 86–7, 126, 137, 138, 146–7, 152

see also children's rhymes
Porter, Cole 69, 138
portmanteau 113
potato
　jacket 89
　shoestring 130
pouch 109, 111, 113
powder room 144
prank 61
pride 20, 23, 37, 58, 64–5, 82, 112
prohibition, blanket 15
promontory 30
protection 50, 133, 144
proverbs 10, 15, 41, 119, 123, 129, 130, 147, 150–1
pub names 71
public, washing dirty linen in 92
public eye, shrouded from 131–2
public purse 113
pucker up 113
pull rug from under me 121
pull your socks up! 136
punishment
　dressing down 33, 58
　hair shirt 126
　putting boot in 19
　with slipper 93
　tan your hide 23
　to carpet 33
puppet, sock 135
purse 4, 25, 109, 111, **113–15**, 136
purser 114
pursuit 141, 142
pyjamas **115**
Pyramus and Thisbe 151

Queensbury, Marquess of 11
quilt 47, 49–50
quotations
　Austen 29–30
　Mark Twain 20
　Rabelais 49
　Shakespeare 14, 20, 40–1, 71, 74, 80, 114, 121, 151
　see also poetry & verse quotations

rabbit, pull out of a hat 84
Rabelais, François 49
race/racing
　dropping a hat 85

flag is down 65
put your shirt on a horse 126
rich purse 114
rag
　paltry 105
　rags to riches 37
　rug 121
ragamuffin 86
Ragnar 'Loðbrok' 23
rain 124, 132
rascal 61
read between the lines 94
Rebeccaites 56
rebooting 19
recklessness 18
red
　carpet 33
　collar worker 44
　flag 47, 65–6
　garter 71
　hat/shoes and no knickers 68
　ribbons 116
　significance 68
religion
　pall 104
　robes 117–18
　veil 150
　see also clergy; monks
restlessness 1
retail **116**
retirement 83–4
Rhodope 129
rhymes see children's rhymes; poetry & verse; popular songs & rhymes;
rhyming slang 31, 37, 54, 84
ribbon **116–17**
　blue **17**
　cross-gartered 71
　filet 63
Richelieu, Cardinal 76
riddles 55, 108, 145
robbers 19, 41, 111, 117, 119, 121
robe 19, **117–18**, 151
Romans 3, 4, 9, 10, 11, 14, 24, 42, 50, 93, 104
Roosevelt, Theodore 58, 64
rosary beads 34
Ross, Lord 75
round rimmers 85
Rousseau, Jean Jacques 24
Royal Charter 74, 78

164

rubbish 105, **118**
ruffer 119
ruffian **118–21**
Ruffiana 119
ruffing 121
ruffle 118, 119
ruffs 6, 118–21
rug 32, 33, **121–2**, 155
rugged 122

sabot 37, 122
sabotage 121, **122**
sack **122–3**
   from work 33
   hitting the 10
   stuffed 47
   to get someone into the 10
sackcloth and ashes **123**
sacrifice 127
sad sack 123
saddles 8, 30, 35, 122
sadness 20, 123, 129, 132
sailing
   bonnet sail 18
   canvas sails 28
   mantle 144
   nail your colours to the mast 47
   sheet 124
   shrouds 132
St Cuthbert 131
St David 80–1
St Nicholas 35
sanitary pad 41, 65, 70
sari 102, 145
sark 51
satchel 122
Saxony, Duke of 148
scamp 61
Scapa Flow 31
scapegoat 63
scapegrace 61
Scarborough Fair 39, 40, 126
scaredy-pants 106
scarf
   bandanna **6**
   kerchief 80
   muffler 100
Scarlet Woman 68
Schulz, Charles 16
secrets
   keep it under your hat 82–3
   keeping up your sleeve 133
   telling 4
   washing dirty linen in public 92
security blanket 16
sewing
   haberdashery **76–8**
   mending/patching 8
   quilting 47, 49–50
   toile 144
   tucking 13
sex/seduction
   euphemisms for 10, 18, 33, 68, 73, 80, 103, 128
   petticoat/pinafore discipline 107
   pillow book 111
Shafto, Bobby 25
Shakespeare, William 14, 20, 40–1, 71, 74, 80, 114, 121, 151
sheep
   mutton dressed as lamb 59
   sheepish 123
   sheep's eyes 123
   wolf in sheep's clothing 40
sheepskin **123–4**
sheets **124**
   folded 1
   ropes 124
   tucking 13
   winding 124, 131
shell, dud 59
shingles 156
ships see liner; navy; sailing
shirt 51, 82, **124–7**, 134
shirttail 126
shoe **127–31**
   brogue **24**
   clogs 37
   filet 63
   glass slipper 72–3
   on the other foot 130
   other people's 129
   vamp 149
   see also boot
shoemakers 128, 130
shoestring/shoelace 130–1
shoon 129
short-sheeted bed 1
shorts, eat my **60–1**
show the flag 64
show must go on 100
shroud 124, **131–2**
Shuckburgh, Dr Richard 53
shut up 11, 135

Signy 134
silk
   bombast 6
   coat of arms 2
   haberdasher 78
   handkerchief 80, 82
   Mantua 97
   padded jacket 31
   sari 102
   stockings 136, 137
   to make a silk purse out of a sow's ear 114–15
   turban 148
   weavers 107
silver lining 94
Simpson, Bart 60
singing
   a cappella 34
   to belt out a tune 11
   see also music; popular songs & rhymes
skirt 42, 55, 90, 109, **132**
skirting board 132
sky, blue **17–18**
slacks 95, 105
slap, sporran 136
sleep
   beds 9–10
   blanket drill 15
   going to 10, 13
sleeve **133–4**
   asbestos 3
   garter 71
sleeveless 134
slip 133
slipper
   glass **72–3**
   smacking with 93
slipshod 130
slovenly 130
smarty-pants 105–6
Smith Miller, Mrs Elizabeth 16
smock/smocking 68, **134–5**
snake 44
snow 15, 43, 124
snuff 143
snug as a bug in a rug 121
Socialist Party, Red Flag 66
sock it to me 135
socks **135–6**
   garters 69
soldiers see army
songs see children's rhymes; popular songs & rhymes

sound, blanketing 15
speak my mind 25
speak off the cuff 48
spoilsport 15
sporran **136**
sport
    baseball mitt 99
    end zone 156
    hat trick 84
    muff a catch 101
sports jacket 90
spying 36
standard, flag 8
star and garter 70–1
stealing 111, 112, 117, 147, 153
sticks, bundle of 9
Stillingfleet, Benjamin 45
stockings **136–9**
    bluestocking 45, 138
    Christmas **35**, 138
    garters 69, 71, 91
    hose 86
    ladders 138
    vamp 149
stocks 137
straight-laced 96
straightjacket 46, **139**
strap/tie 63, 116, 130
stroller 90
Strong, Barrett 84
stuffed shirt 126, 141
stuffing 8, 47, 49, 50, 60
submission 65, 123, 145
subtle **139**
Sufi tradition 46, 117
suit **140–2**
    of armour 2
    birthday 14, 101
    handkerchief 82
    jacket 89–90
suitcase 113, 142
suitor 142
Sumptuary Laws 99, 118–19, 120
sunset/sunrise 15
superstition 75, 107, 127–8, 130, 133, 138, 149
sure thing 35
suspenders 21, 136, 137
swaddle **142**
swag/swagman 152, 153
sweater 90
sweater girl 91

sword
    bastard 8
    and buckler 24, 25
    length 119
    to be breeched 24

tablecloth 2, 3, 32, 144
tailcoat 42, 68, 89
tailors 10, 41, 58, 90, 108, 116, 119
take my hat off to you 82
take off the gloves 73
talk/talking
    pillow talk 110
    put a sock in it 135
    skirting round a topic 132
    sock it to me 135
    speak my mind 25
    speak off the cuff 48
    talk through your hat 83
tanning hide 23, 93
Tartuffe 126
task
    laundry list 92
    tiresome 55
    to buckle down to 25
tassel 101, 128, 136, 143
Tatterson, John 69
teddy 26
teenage delinquent 86
telephone line 94
television programmes 54, 59, 66
telling off 19, 33, 59, 99, 155
tents, canvas 28
Thatcher, Margaret 79
theatre
    curtains 49
    flag is down 65
    men wearing women's clothing 55
    motley crew 100
    mummers 54
    show must go on 100
thieves 6, 19, 41, 111, 112, 117, 119, 121, 147
thinking
    blue sky 18
    cap 29
    put a wet towel round your head 145
thread 63, 115, 119
throat, asbestos 3
throw hat into the ring 85

throw in the towel 145
tidy 57
tie
    and jacket 90
    and tails 42, 68
tie/strap 63, 116, 130
tip 76
tire 63
titfer 84
toff **143**
toffee-nosed 143
togs **143**
toil/toile 144
toilet **143–4**
tongue twister 80
top brass 85
topcoat 90
torn to ribbons 116
tout de suite 141–2
towel **144–5**
town and gown **145**
trade, retail **116**
train, flagging down 65
train-spotters 1
tramp 5, 59
transvestite 89, 152
travesty **145–6**
trousers **146–7**
    all mouth and 83, 147
    braces/suspenders 21
    breeches/britches **22–4**
    caught with trousers down 147
    enough blue sky to make 17
    flying by the seat of 66
    knickerbockers 92
    pants 105
    slacks 95, 105
    which side do you dress? 58
truth, naked 102
tuck 12–13
tucked up 13
tucker **11–13**
tuckered out 13
tucking 13
tuft 143
tulip 148
tune, to belt out 11
turban **148**
turncoat **148–9**
tuxedo 90
Twain, Mark 20
twinset and pearls **149**

undercoat 42, 90
underwear
   bikini 14
   bloomers 16
   brassiere/bra **21–2**
   busk 95, 96
   camiknickers **26–7**
   clouts 41
   corset 16, 18, 21, 24
   eat my shorts **60–1**
   going commando 102
   knickers 26, 68, 91
   loose jumps 90–1
   pants **105–6**
   petticoat 42, **106–8**
   sark 51
   vest 16, **151–2**, 156
unfrocking clergy 67
uniform 32, 45, 53, 78, 87, 94, 101, 116, 141
urban legends 3, 44, 122

vamp 149, 150
vamping a tune **149–50**
veil 104, 139, **150–1**
vellum see parchment
velvet 74, 117
Venus, pocket 112
Verrazano, Giovani da 61
verse see children's rhymes; poetry & verse; popular songs & rhymes;
vest 16, 87, **151–2**, 156
Vesta matches 152
vestal virgins 152
vestibule 151
vestment 151
vestry 151
Vicar, William Lee 137
Vidar 130
Vikings 22, 23, 134
Virgin Mary 18, 32
voile 150
Völsunga Saga 134

votes, canvassing 28

waistcoat 151
Wales
   Mabinogion 75
   Rebeccaites 56
   Welsh hats **153–4**
Waltzing Matilda **152–3**
wardrobe 118
washcloth 143, 144
washing, laundry **92**
washroom 144
wasp 146
watch, escapement 61
weather
   blue sky **17–18**
   cloud/mist 86, 150
   rain 124, 132
   sheet lightning 124
   snow 15, 43, 124
   sunset/sunrise 15
weaving, blankets 15
Wee Jack 5
well-dressed 59
well-heeled/shod 130
Welsh hats **153–4**
wet blanket 15
white coats, men in 46
white collar worker 43–4, 71, 85
white flag 65
white glove 75
white monks 45
white as a sheet 124
Whitfield, Norman 84
Whiting, George 58
whoopee cushion 50
wigs 144, **155**
wind
   leave you blowing in 92
   throw you cap into the 30
winding sheet 124, 131
windmill, throw your bonnet over 18

window-dressing 58
wipe the floor with someone 99
wison 53
witchcraft 51, 69, 75, 150
wolf in sheep's clothing 40
women
   bit of skirt 132
   blowzy 17
   completing her toilet 143, 144
   curtain lecture 49
   does my bum look big in this? **54–5**
   female tramp 5
   feminists 22
   loose women **95–6**
   men dressed as 55–7, 89, 107, 152
   pocket Venus 112
   robed in beauty 117
   as soldiers 56–7
   sporran basher 136
   vamp 150
   wearing trousers 147
wooden overcoat **156**
woollen
   motley fabric 99
   Sufi clothing 46
   to lie in the 14
work
   clothing 85, 141, 142
   duvet day 60
   giving the sack 122
   retirement 83–4
   roll up your sleeves 133
   step into someone's shoes 129
   toil 144
   wearing two hats 84
writing desk 25–6

yarn, loopy 24
yellow flag 65

zone **156**

Here's what they said about Elinor Kapp's first book **Rigmaroles and Ragamuffins**:

"Elinor Kapp is an extraordinary woman. Not content with being a medical doctor and a psychiatrist, she has become a textile artist. She recently wrote a book called **Rigmaroles and Ragamuffins** ... this is a fascinating journey through a world of materials, tools and manufacturing processes, on the way shedding light on the forgotten origin of many words now part of the English language."
Diarmid O Muirithe, in The Oldie (Feb 2009)

"a gem of a book".
Let's Knit Magazine (Dec/Jan 2007)

"... a fascinating insight into the way textile-related words and phrases have made their way into common usage. Many of them you will be familiar with ... but some will amaze you."   *Classic Stitches (No 84)*

Classic Stitches magazine made **Rigmaroles and Ragamuffins** a prize in their Christmas Competition: "If you are looking for a book that is a bit different, perhaps seeking a gift for a stitcher who also loves reading the English language then this is the perfect choice."

"English has the largest and most motley vocabulary in the chattering world. Textiles provide one of its richest seams. Elinor Kapp has written a delightful guide through its semantic maze."   *Philip Howard, of The Times*

* * * * *

Rigmaroles and Ragamuffins was published in 2007 and re-issued in a second edition in 2013. It is available anywhere and everywhere that books are sold.
ISBN 978-0-9574759-0-8

Lightning Source UK Ltd.
Milton Keynes UK
UKOW05f0826270916

283894UK00002B/99/P